O9-AIF-517

Agrarianism and the Good Society

Culture of the Land

A Series in the New Agrarianism

This series will be devoted to the exploration and articulation of a new agrarianism that considers the health of habitats and human communities together. It will demonstrate how agrarian insights and responsibilities can be worked out in diverse fields of learning and living: history, science, art, politics, economics, literature, philosophy, religion, urban planning, education, and public policy. Agrarianism is a comprehensive worldview that appreciates the intimate and practical connections that exist between humans and the earth. It stands as our most promising alternative to the unsustainable and destructive ways of current global, industrial, and consumer culture.

AGRARIANISM
and the Good Society

Land, Culture,
Conflict, and Hope

ERIC T. FREYFOGLE

THE UNIVERSITY PRESS OF KENTUCKY

Publication of this volume was made possible in part by
a grant from the National Endowment for the Humanities.

Scholarly publisher for the Commonwealth,
serving Bellarmine University, Berea College, Centre
College of Kentucky, Eastern Kentucky University,
The Filson Historical Society, Georgetown College,
Kentucky Historical Society, Kentucky State University,
Morehead State University, Murray State University,
Northern Kentucky University, Transylvania University,
University of Kentucky, University of Louisville,
and Western Kentucky University.

Editorial and Sales Offices: The University Press of Kentucky
663 South Limestone Street, Lexington, Kentucky 40508-4008
www.kentuckypress.com

11 10 09 08 07 5 4 3 2 1

Library of Congress Cataloging-in-Publication Data

Freyfogle, Eric T.
Agrarianism and the good society : land, culture, conflict, and hope /
Eric T. Freyfogle.
p. cm. — (Culture of the land)
Includes bibliographical references and index.
ISBN-13: 978-0-8131-2439-1 (hardcover : alk. paper)
ISBN-10: 0-8131-2439-5 (hardcover : alk. paper)
1. Land us—Environmental aspects—United States. 2. Environmental
degradation—United States. 3. Environmental ethics—United States.
4. Human ecology—United States. 5. Social values—United States.
6. National characteristics, American. I. Title.
HD205.F73 2007
306.3'490973—dc22
2006036947

This book is printed on acid-free recycled paper meeting
the requirements of the American National Standard
for Permanence in Paper for Printed Library Materials.

Manufactured in the United States of America.

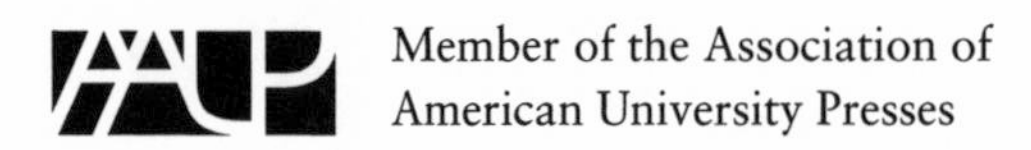

Contents

INTRODUCTION

One way to judge a people is to look at the ways they use nature—the land, broadly defined to include its soils, rocks, waters, plants, animals, and sustaining processes. A culture writes its name on land for all to see. Is the soil kept fertile and in place? Are waterways clean and full of life? Are tracts of land devoted to uses for which they are ecologically well suited? Are landscapes sensibly laid out and pleasing to the eye and ear? And are the modes of living and working on land likely to endure for centuries, without nature lashing back? The land reflects not just what people have done but who they are, what they understand, what they value, and what they dream.

Good land use is challenging enough on a single land parcel, tended by a single steward. Those who have tried the task know what it entails, particularly when land is ecologically sensitive. The challenge expands when we shift scales and talk about good land use over entire landscapes. Going further (as we need to), let us assume that the dozens or hundreds of people tending these lands need to make a decent living. Economics is important, and so the inquiry broadens further. Land use does not take place in isolation, socially, economically, or ecologically. It occurs within a social order that surrounds and includes the land and land manager. And so we bring in contemporary culture, broadly conceived to cover economic relations and political structures as well as our values, loyalties, affections, and institutions.

Good land use is likely to occur—not merely on parcels here and there but everywhere—only when it stands as a shared aspiration. Society must embrace the goal and work toward it. This need for social support has always been with us, but now it is urgent. Land-use practices are increasingly guided by global economic forces. Degradation unfolds at broader and broader scales. Dead zones grow in the Gulf of Mexico because of land uses hundreds of miles away. Forested mountaintops in Appalachia are shoved aside to extract coal for distant power plants. Meanwhile, expansive farm fields in the corn belt are stripped bare of all visible life—they are hardly more diverse biologically than parking lots—by owners who respond to global price signals that ignore all local considerations. To get good land use, it is not enough to educate capable farmers and foresters; good culture is also necessary. And good culture does not come easily.

Good culture would honor land and the wise use of it. Necessarily it would honor the skills and labor required to bring it about. This would be, literally, an agriculture, from Greek and Latin roots meaning field, land, and land cultivation; a form of agrarianism centered on the widespread use and good tending of all types of land.

What kind of culture is needed to take care of land, given the economic and political realities of our day? What elements of modern culture require challenge and alteration if good land use is to become our ideal and duty? What might a land-respecting politics look like, and what kinds of leaders would we need to place good land use on the national agenda? These are the questions I take up, in an effort to envision a culture that is well rooted and planning to stay put.

When talking about land-related ailments—soil erosion, urban sprawl, sagging farms, deteriorating neighborhoods, wildlife habitat loss—our tendency is to refer to them as environmental problems. The implication is that the earth is somehow at fault for them. But the true causes are rooted in human behavior, as we know in our quiet moments. Environmental problems arise because we are not living well on land.

Our responses to our land-related ills tend to take two forms (aside from mere denial). The less common response is for individuals to detach themselves from the system that produces the ailments, to live apart from the destructive practices of the age. Get back to the land, disconnect from the electric grid, and cut unnecessary ties with the industrial juggernaut. More frequent than this simple-living response are the various efforts made to diminish our system's bad consequences by working within it. The activism here typically accepts the dominant language and modes of thought of the day. It embraces the market, creates few cultural waves, and is often content merely to preserve a land parcel here and there.

These are the options as most people see things: either drop out, so far as one reasonably can, or work within the mainstream and settle for small gains.

These options are becoming less attractive as our landscapes slide down, ecologically and aesthetically, and as the forces resisting good land use gain strength. The one is largely futile because it entails escape. The other dispenses Band-Aids to keep the industrial order hobbling along. Neither does much to remedy the root causes of decline.

My aim is to clarify a middle course that does offer hope. I do not mean a course that wanders somewhere between the extremes of unrestrained industrial land use and the exaltation of wild places. That middle ground we know about: it is too vast and formless to be of help. Instead, I mean a course that identifies sound cultural values and that seeks to make these values dominant. A well-guided middle course would begin within the modern order but push for substantial change where it is needed most—at the levels of perception, language, value, and motive. This middle course would foster new ways of seeing and valuing the land. It would recognize our interdependence, with one another and with nature, while showing awareness of human ignorance. It would commit to lasting ecological health while remaining mindful that only a sound culture can use nature well.

In these pages I criticize much environmental work taking place in the United States, though reluctantly and with respect for the labors involved. Too often environmental efforts are poorly aimed.

They draw attention to the visible symptoms of land sickness but then fail to trace that sickness to its social and cultural roots. Treatments are localized. Open sores are merely covered up. Meanwhile the disease rages on. As for those who push for change at the level of individual behavior—whether dropping out or changing how we spend money—there is much to admire there as well. Certainly we should support local farmers and "green" products. Of course individuals can make a difference. But ascetic living is a hard, slow sell, and the salespeople on the other side are better funded. In any event, the system itself is a big part of the problem. We simply must confront it. Individuals acting alone cannot deal with many land-use deficiencies. The forces of resistance are too massive. The system itself requires change.

Ecology will play a central role in this reconstruction of modern culture, given the wisdom of aligning with nature's processes. To keep all the biological parts is also wise, even ethically compelling. And we have sound reasons to keep wildness near at hand. Still, our primary focus needs to fix on the lands where people live and work. What does it mean to live well on land, in terms of actual land practices? What kinds of perceptions and values would encourage good land use? What can we expect of one another by way of responsible behavior when it comes to tending land? Yes, we need to honor the individual, but in what ways? Yes, we need to promote civil liberty, but of what type—positive or negative, individual or collective? For generations the moral call has gone out to love one another. But what can love mean in an era when the land-use decisions of a single person can drag down an entire landscape?

In the opening chapters, I look to nature as the beginning place to forge a cultural order that respects land, people, and the bonds between them. Chapter 1 considers the natural areas that dot our landscapes—our parks, wildlife refuges, and forest preserves. These lands provide homes to wildlife and allow nature's processes to operate more freely. In an essay that began life as a talk to land-management professionals, I explore the cultural values that natural areas embody and the ways they might become centers of cultural change. To make that happen, we need to conceive of our natural areas not as places of escape—as wild enclaves set apart—

but instead as integral elements of larger landscapes. In what ways can wild places bring health to the lands and the people who live around them? How can we think and talk about these life-filled places so that they become a form of cultural leaven? This is the central challenge.

Chapter 2 continues the inquiry by revisiting some important writings by conservation giant Aldo Leopold, an early advocate for wilderness and scientific wildlife management. Leopold marshaled varied arguments in favor of wilderness, many dealing with the values wilderness had for recreation and wildlife. Less remembered are the culturally centered arguments that dominated his late conservation thought. Leopold worried endlessly about the challenge of getting people to live well on land—the "oldest task in human history," he termed it. Good land use, he knew, could arise only from good culture. To get better land use, we would need better culture. This is the part of the wilderness protection story that we have overlooked. Wilderness, Leopold said, could serve as a place "to organize yet another search for a durable scale of values." Wild places could become—we need them to become—places where we can stand and look out to identify better modes of life.

Chapter 3 turns to land of a different type—the homestead farm. Again the focus is on the cultural side of land tending. To frame matters, I draw upon an extraordinary novel by Charles Frazier, *Cold Mountain*, set in western North Carolina late in the Civil War. In Frazier's story of two women, Ada and Ruby, we witness how nature and culture might come together on a small farm, giving rise not just to health and stability for the two women but to alluring visions of a rural society well settled in its natural home. By extension, the story illuminates how an attentiveness to land and an affection for a particular place might transform the ways we see the world and our place in it.

Chapter 4 puts these observations into practical frames by challenging the simplistic ways we typically talk about our current choices on land-use issues. The conventional dichotomies we use are both familiar and unappealing: Do we worship nature or pay attention to people? Do we take the liberal route or the conservative one? Do we honor individual freedom or engage in public coercion? These familiar dichotomies, I contend, all reflect poor thought. To shed better light on our cultural choices, I identify seven other dichotomous options. From varied angles they frame more clearly the forks in the cultural road ahead. To identify these choices is to see more clearly how we are now living and how things might change.

The next two chapters have to do with private property and private rights in land. Property is a social and cultural institution as well as a legal one. It is the social and cultural side that draws attention here—the culture of owning, and how owning land necessarily entails belonging to place. Chapter 5 explores the ways that private property works, paying particular attention to its social roles. As a way of managing land uses, private property is a surprisingly flexible institution; it can define and assign land-use rights in a wide variety of ways (not just the ways we know). I explore the institution and our dominant culture of owning by looking at the presumed chasm between publicly owned lands and privately owned ones—an issue practically important in the American West and culturally influential everywhere. In fact, the two forms of ownership are quite similar. Public and private interests are implicated on all lands, regardless of nominal ownership. To see this is to raise the possibility of crafting new and better ways of mixing public and private everywhere—through carefully tailored private use rights and through better designed land-management regimes. Chapter 6 continues this digging by comparing our evolving thoughts about private property with our similarly evolving thoughts in two related fields of knowledge—ethics and ecology. It is useful to compare these three cognate fields because the overarching trends within them are similar. In each, an early commitment to an organic view of the world gave way to greater attention to individualism and autonomy. Context and community counted for less; abstract reasoning and atomism counted for more. In all three fields, this shift toward abstraction and individualism brought vast gains—secure private rights, economic growth, greater respect for people, more productive lands. But in all three the shift went too far and has stimulated a counter-shift back toward the intellectual reassembly of the fragmented social and natural parts. For various reasons this counter-move toward community needs to continue. As it does, however, we need to use caution, because many forms of community building and reassembly can cause more harm than good.

The final three chapters turn to politics and the challenges of working together to resettle our continent. Chapter 7 draws upon an important novel, *Jayber Crow*, by perhaps our most prophetic voice on nature and culture, the Kentucky writer and farmer Wendell Berry. It questions Berry's implicit claim in his book that the key to moral growth is to love one another. In the protagonist of this masterpiece, set in a rural Kentucky landscape under attack, we spot the strengths and weaknesses of American-style agrarian culture. The chief weakness is the failure of Crow and his like-minded agrarian neighbors to get together as citizens to protect their home. Not more love but more democracy is what they need, and what we now need, to live better on land. Like private property, democracy builds upon cultural values, particularly the tradeoffs we make between individual autonomy and secure social bonds. The failings of Berry's fictional agrarians continue to plague our real-word landscapes. We need thoughtful citizen-leaders who can identify a sound path ahead and help move people down it. Chapter 8 takes up that need using an unusual format—a job advertisement for an ideal environmental leader. Chapter 9 brings things to a close by commenting on the political scene. Why have environmental ills largely disappeared in recent national elections? And what might be done to get them back in? To revive these issues politically, friends of the land need new rhetoric. To get that rhetoric, we need to think more clearly and ambitiously about what we are trying to accomplish.

Imagine a typical high-tech farm field, American style. Massive graders have flattened this field, making it accessible by the biggest industrial machines. Fencerows are long gone. So are the old farm creeks and ponds. The land manager begins the new year by killing every living thing on the land except microscopic organisms. A single species of life is then introduced, perhaps a bioengineered wonder quite different from any wild plant. The species probably cannot survive without constant human attention. It might be sterile. The plant's seed or fruit has been engineered for appearance, "shipability," and ease of harvesting. There is little concern about nutrition or ecological effects. Chemicals are deployed to keep other life forms at bay. "Farmers" interact with this land only from cocoons atop noisy machines. Key decisions are made in offices far away. If a human hand touches the resulting produce, it is likely that of a low-paid worker on an assembly line that washes, sorts, and packs. Her building may have no windows. She has probably never seen the field that produces the crop. The people who buy the produce will know nothing more.

We can think of this farm scene as a contemporary counterpoint to a medieval cathedral. More than a place and human construct, the field is a cultural emblem of who we are and of how we understand our place. It is as value laden as any stone edifice. That the values embedded in the field now enjoy influence we must readily admit. Yet how many people know that such fields exist? How many recognize the clash between this industrial scene and any sane notion of healthy lands and healthy communities?

The facts are out there, waiting to be heard. What we lack are good ways to talk about them. Facts float in the air until we have places to put them—frameworks that infuse the facts with meaning, cultural structures that allow us to pass judgment. Here is where a sound agriculture comes in: to give the world meaning, to situate us within it, to help us know how to live.

1

Life in the Enclaves

The work of managing a natural area—a wildlife refuge, park, public forest, wilderness reserve—does not fit easily into American culture. When well done, the profession entails ways of thinking, valuing, and acting that stand culturally apart. It deviates from, and indeed calls into question, much of what America is about.

For starters, the work requires a long-term perspective. It means thinking about and planning for the very long run. This perspective contrasts with the short-term attitude that characterizes much of the modern age, whether it is the business that plans quarter to quarter, the student who looks ahead to the next test, or the worker who lives paycheck to paycheck. Long term is not the norm. Then there is the fact that managers are interested not in what they take from the land but in what they add to it. In the dominant culture, land use means extracting, cutting, and reshaping. In natural areas, managers labor to nurture, grow, and protect. They succeed by giving, not by grabbing.

In the dominant culture, progress is measured in dollar terms. Not so in natural areas, and not so in the minds of natural-area managers. Their success comes as their lands gain in biological diversity and health. Healthy lands in fact are more valuable than sick ones, and we might crudely gauge that value in dollar terms. But the confident land manager knows better than to think this way. True gain is measured by the signs of natural health—by the spongy forest soil and by the rare ducks that have fledged—not by what calculators say. In dominant culture it is the age of supersize it, of Hummers, McMansions, and tottering charge-card debt. The natural area, meanwhile, is a place and a culture of caution and restraint. It is where we control our appetites, where we resist the temptation to sever resources and cash them in now.

Perhaps above all, the natural area and the natural-area manager stand opposed to the fragmentation that so dominates our age: our tendency to fragment landscapes into discrete parcels and political units, to divide time into brief units, to separate land-use activities and allocate one activity to each land parcel, and to fragment lines of intellectual inquiry so that each "expert" is assigned a specialty and expected to keep to that specialty. The natural area and thus the natural-area manager are about transcending these boundaries. The components of a natural area form a complex, seamless whole, not a collection of parts. Birth, growth, and death unfold in overlapping cycles, defying boundaries of time and space. A farm field planted in corn yields a single commodity that arrives in annual increments. The natural area, in contrast, produces yields that flow and turn like a river, ceaselessly coming and going. The land manager in her work must bring to bear a range of skills and understandings. Narrow expertise simply is not enough. The scientific, social, ethical, political: all must come together, on the ground, to do the job well.

Natural areas also stand out because they provide few jobs and no places for people to live. In the dominant culture, lands are most valued when people use them intensively. Human activities count, particularly focused, repetitive activities—the retail store that is open around the clock, the high-rise office building, the ubiquitous parking lot. Natural areas embody a different scale of values. Diversity counts above specialization; nonhuman life stands above human activity; the rare and unexpected occurrence brings particular delight.

And then there is the conflict between natural-area protection and the sweep of American history. America forged its identity on the frontier. It was a culture whose parts came together by taming and reshaping land. To frontier Americans, John Locke's labor theory of land use made sense: Nature held value only insofar as humans invested labor to change it. Unaltered nature possessed little worth. Natural areas upend this ideology of conquest, or they mean to. Frontier Americans changed prairies into farms. Natural-area managers perform the harder work of returning farms to prairies. For them, nature standing alone can have great value. Altering land may or may not bring gain. The ecological restoration that takes place in natural areas directly challenges the successes of our cultural ancestors who plowed the plains and were proud of it. Indeed, to restore nature in that way can seem un-American. When restoring wolves is labeled success, we inevitably call into question the wolf killing of generations past.

If managers feel out of step in today's world, there are good reasons for it.

Much as natural-area managers and restorationists are out of step culturally, so too the natural areas themselves are ecologically and culturally adrift. They float in a world of motion. They are adrift in nature in that the ecological forces sweeping over landscapes enter and pass through them, always leaving them different. Nature is inherently dynamic, and natural areas are part of that dynamism. Winds, floods, plant and animal migrations—all are at work, bringing ceaseless change. Natural areas are adrift culturally in the sweep of human history. A particular landscape is "natural" only in comparison with other places that people have altered more. Indeed, the whole idea of a natural area is a human creation, and inevitably so. People decide what qualifies as natural, whether and why natural areas deserve protection, and how such places will be run. A natural area is surrounded by a complex culture, which is just as dynamic as the ecological forces at work: different a generation ago, different a generation hence. The reality, in short, is that change never ceases, ecologically or culturally. To retreat to a natural area is thus not to escape change. It is akin to climbing aboard a timber in a free-flowing river and getting carried by the many currents.

Professionals who work in natural areas are not in nature just for the ride, of course. It is their calling to keep the timbers afloat and sensibly directed. Much of the change unfolding around them, in nature and culture, they cannot and should not resist. But it is just as foolish to get tossed about passively, giving in to every wind of change. Exotic species, excessive deer, annoying landowner-neighbors, politicians who know only jobs and tax base—these and other forces can demand land managers' resistance. To manage nature you need to keep your wits about you—going along with change when it is wise to do so, resisting when it is not, and striving always to keep nature intact for new generations.

A natural area can function well only if it is integrating into a larger physical landscape. It needs to contribute to the sound functioning of a bigger nature. It also needs integrating into the surrounding cultural landscape. Thus managers constantly need to find effective ways to explain to people why their natural areas make good sense. As nature shifts ecologically, the physical management of the natural area ought to change accordingly. So too, as surrounding culture shifts, the natural area as a cultural artifact needs resituating, in terms of the ways it contributes to surrounding culture. This means, for the land manager, having continuously to sell and resell the natural area to neighbors and other constituencies. The manager must refine and refocus the various rationales for keeping natural areas intact.

Natural areas do provide benefits for people who live on surrounding landscapes, and there are many ways of talking about those benefits, some more lucid and persuasive than others. Sometimes the sales job can stress how natural areas give people what they want. Sometimes it should convince people to want something they have not yet learned to want. Likewise, good advocacy can build upon what people already see. Other times it will call them to see things anew.

This labor of raising public support for nature is hardly easy. It never has been. American culture in particular poses special challenges, even though many Americans are fond of wild creatures and picturesque landscapes. The special challenges arise not only because people live in urban areas, cut off from well-functioning nature. And they arise not merely because, as David Orr observes, the average young person knows perhaps fifty corporate logos for each native species he can identify; as Orr phrases it, today's youth are autistic ecologically.[1] The alien-ness of contemporary culture runs deeper than this. Three strands of it are particularly troubling.

One disturbing cultural component is the capitalist or market mentality that dominates how we see and interpret the world. I mean by this our tendency to fragment everything into pieces and then to value the pieces in the market (or by using market substitutes). Nature is typically fragmented into distinct parts—into parcels of land and discrete natural resources—its parts then valued piece by piece. Humans are fragmented, too, our lives divided into specific roles and our labor treated as another market commodity. It is hard to spot and appreciate the connections when all we see are the pieces. It is hard to uphold the long term when the market encourages consumption today.

Related to this capitalist mentality and equally troubling is the embedded individualism of the modern age. We view people as autonomous individuals, honor them as such, extend them rights, and then give them wide berth in their personal preferences and moral judgments. Gaining ground around the world, this liberal orientation is perhaps strongest in the United States. Here, countless social issues get framed as conflicts over individual rights, rather than (as they are elsewhere) as problems confronting society as a whole. Individualism has its virtues, to be sure. But a virtue carried too far can become a vice. Moral relativity is one consequence of individualism taken too far. So too is the contention sometimes made in land-use debates that there are in fact many land ethics, not just one. The subtext here, apparently, is that one land ethic is nearly as good as another. Implicitly, land users ought to be free to choose a land ethic as they might choose a suit of clothes. As individualism of this type continues to gain strength, the community as such counts for less and less. In too many ways, Americans have become simply shifting collections of individuals, weakly linked and shallowly rooted in place.

The third element of culture that challenges nature preservation is the intellectual fragmentation of the modern age, our tendency to divide knowledge into tiny bits and to reward people who specialize in one or a few of the bits. The result: we wind up with many people knowing much about small subjects and few people who possess broad understanding. Analysis is the skill of the day, whereas good synthesis is little honored and is hard to identify when we see it. In specialty tasks we often do well, but we stumble when it comes to engaging critically with the larger structures of thought and culture, with the kind of wide-ranging thought we need in order to identify our rightful place in nature.

Nowhere is this latter problem more evident than within the modern academy, where specialists abound but where it is increasingly hard to find scholars who integrate well at larger scales. Too few academics are skilled at drawing together knowledge from a variety of intellectual perspectives. Particularly lacking are citizen-scholars who can go beyond integrating sciences to draw in history, philosophy, social thought, and serious cultural criticism. The modern academy, seeing this problem of intellectual fragmentation, responds by assembling various specialists and labeling them an "interdisciplinary" team. But an assembly of specialists too often remains just that, a collection of specialists. The wise and critical integration of knowledge requires a much different kind of scholar, one who can critically examine all the parts and then bring them together wisely.

For the natural-area manager, these cultural elements are all distinctly troubling, despite their virtues in moderation: the capitalist, market mentality, which fragments nature and views everything and everybody as a commodity; the ardent individualism and rights-based reasoning of the age, which makes all talk of community suspect; and our intellectual fragmentation, which divides knowledge into bits and loses much in the process.

These conflicts with modern culture arise, of course, because natural areas are not collections of resources. They are integrated, ecological wholes. Their benefits compute poorly in market terms. They are particularly undervalued when subjected to discount rates that reduce the far future to irrelevance. Many of these benefits, moreover, extend to the community as a whole, to everyone who lives in the region, whether or not they ever visit. Benefit calculations that dwell on visitor days miss them entirely. Natural areas, of course, do benefit hunters and birders as individuals, but these direct benefits to visitors compose only a part of the benefit package. And as for managing the lands, it simply cannot be done by a person expert in one profession. It is not enough for a manager to know one group of animals or plants or to understand only hydrologic flows, soil types, or forest management. Conservation writer Aldo Leopold criticized this fragmented approach in 1934, a time when New Deal leaders were assigning the nation's many conservation challenges to different agencies: "The plain lesson is that to be a practitioner of conservation on a piece of land takes more brains, and a wider range of sympathy, forethought, and experience, than to be a specialized forester, game manger, range manager, or erosion expert in a college or a conservation bureau. Integration is easy on paper, but a lot more important and more difficult in the field."[2] Nature is an integrated whole, and good managers treat it that way.

Ultimately it is an alien world that surrounds the natural area, not hostile but distinctly suspicious. People want to be shown the benefits of putting nature off-limits. And too often they arrive in the hearing chamber guided by values and assumptions that tilt in the wrong direction.

These American cultural traits have long been around. So have criticisms of them. In the 1930s and 1940s, Leopold complained bitterly about academic fragmentation and its resulting land-use ills, or as bitterly as he could without alienating his faculty colleagues at the University of Wisconsin. Before that, ecologist Victor Shelford of the University of Illinois—another land-grant school—unleashed similar sharp barbs.

This suspicious culture poses stumbling blocks for people who care deeply about nature and who know how it ought to function. In some manner they must come to terms with modern culture. They must decide which cultural elements to accept and which ones to resist. Probably the key challenge is posed by the exceptional influence of individualism—what historian Louis Hartz so famously termed America's liberal tradition. Should nature advocates give in to this cultural trait and adopt the language of individualism, in some way using it to explain why nature is important? Or should they instead remain faithful to the intellectual core of conservation thought, which is foremost about the well-being of life as an integrated community—about the health and welfare of ecological wholes, not about pleasing people as isolated individuals?

This dilemma presents a high-stakes game, and there is no clear answer to it, applicable to all environmental work. In the heated aftermath of *Silent Spring*, Rachel Carson embraced the language of liberal individualism when she testified before Congress about pesticide-use practices. The rhetoric Carson chose, the rhetoric she thought most persuasive, was the language of individual rights. Dangerous pesticides were being applied widely without adequate thought and study, Carson asserted. Pesticides were being sprayed on people's lands and homes, even directly on their bodies, without their informed consent. Reckless pesticide spraying was violating individual rights.

This individual-rights approach has waxed and waned as a rhetorical strategy for addressing environmental ills. In classical liberal thought, environmental degradation is wrong because it interferes with individual liberties or otherwise violates individual rights. This liberal reasoning shows up from time to time in proposals to add a "green" amendment to the federal Bill of Rights. The idea is to create some sort of provision that would protect the environment by enshrining it as an individual right. Many states already have such a constitutional guarantee, typically phrased as an individual right to a healthful or healthy environment. (The right is sometimes matched with a duty to help maintain such an environment.)

Individual-rights rhetoric offers promise in addressing problems such as the ones Rachel Carson had in mind—toxic contamination and air and water pollution. Harmful chemicals can directly insult the bodies of individual citizens. Individual integrity, of course, is a core right. Rights rhetoric like this works much less well in dealing with land-related issues: when it is used to explain, for instance, why we need wildlife refuges, restored savannahs, or wilderness tracts. Illinois activist Joe Glisson found this out in the late 1990s, when he ultimately failed in his legal effort to block a reservoir project that would flood his riverside home and destroy critical habitat of a state endangered species. In Glisson's view, the reservoir impaired his right to a healthful environment under the Illinois constitution. The Illinois Supreme Court thought otherwise. As the court saw things, "healthful" referred only to the health of individual humans as autonomous beings. It did not extend to the health of the land. Human health was legally diminished only by pollution and toxics, not by ecological decline. Health apparently had nothing to do with the presence or absence of other life forms.[3]

The benefits of natural areas do not translate readily into this kind of individual-rights rhetoric. Natural areas help sustain landscapes. They promote wildlife, flowing rivers, fertile soils, and clean air not for humans in isolation but for entire communities of people. Because of this, defenders of natural areas would do better relying on community-based rhetoric. Ecological degradation afflicts us collectively, not as individuals: that is the core message. It degrades integrated communities of life.

Many of the benefits of natural areas spread widely, rippling through natural systems. Benefits extend beyond identifiable individuals to reach future generations, nonhuman life, and ecological communities as such. They include gains that we all enjoy, whether or not we ever enter the areas.

Yet another problem with using individual-rights rhetoric to talk about land is the challenge of finding clear words to express the right. The usual rhetorical form for constitutional rights—a right to be free of some type of interference—simply does not work. The normal reaction is to turn then to ecology for guidance, somehow phrasing the individual right in terms of the land's ecological functioning. This approach, though, stumbles because of the sizeable limits on ecological science. No science, ecology included, can tell us how we ought to live on land. It cannot tell us whether land is or is not in "good" condition.[4] To make such a judgment, more than science is required. Nature in fact gives us considerable room in deciding how we want our landscapes to function and appear, without sapping the land's long-term fertility. Given this flexibility in nature, and given the inherent interconnection of landscapes, good land use can make sense only as a communal goal.

If natural areas are going to thrive—staying afloat in the shifting flow of culture—their defenders need to resist the individualism and fragmentation of the age. They need to challenge key elements of modern culture, the elements that focus on the short term, that fragment nature into pieces, that look to the market for values, and that divide knowledge into specialty pieces, valuing the parts and discounting the larger whole. They need to fight against the notion, ruthlessly embraced by the market, that public policy should merely give people what they want as autonomous individuals.

By necessity, then, managers of natural areas are enmeshed in cultural fights. Their modes of thought and the values embedded in their lands collide with key elements of modern culture. This conflict puts the natural areas at risk. At the same time, it poses opportunities to bring about important change. This is the positive side. Natural areas can serve as places to stand to criticize the modern age. They can become agents and places to promote healthier lands and a more land-respecting culture.

Natural areas are uniquely valuable in reminding us how nature functions ecologically when it is healthy. Unchanged nature, to be sure, cannot sensibly serve as a benchmark for measuring how well we are living on land. On these points we should be clear: Raw wilderness is no place to live. Change is essential. But change can take many forms, more or less respectful of nature's processes. To distinguish good change from bad change requires a benchmark of some sort, a normative standard for evaluating land-use conditions. Many factors will be relevant in piecing together such a standard. One key factor, surely, is whether the land remains productive, fertile, and biologically diverse. Natural areas can help give content to this ideal. Research within them can show us how nature's basic processes work and how we can promote them. By studying in natural areas, we can identify the functional elements of nature that we ought to be respecting everywhere.

Good land use is an elusive concept. We rarely talk about it directly. Once we do raise it, we quickly see how complex it is.[5] Human utility is a key component of it. Utility in turn has numerous aspects, direct and indirect. Many utility-related factors relate to the quality of life rather than the quantity of resource flows.

Ethical and religious values also come into play here. Good land use would avoid taking risks with nature that we cannot afford to lose, and it would recognize our vast ignorance about nature by interjecting elements of caution in our decision-making processes. Although good land use would certainly draw upon science, we cannot reasonably expect science alone to tell us to live. Science is a body of knowledge about nature and a tool for gaining more knowledge. It falls far short, though, of including all the elements we need to decide how we ought to live on land. To make such judgments, we need to bring in a variety of nonscience considerations. Integration is required, and it is tough work.

One place where we ought to find good thought about land use is within the conservation movement. For years, conservation and environmental groups have pushed us to improve our dealings with nature. But the overall movement, sadly, is hardly less fragmented than the academy. Its strengths are matched by weaknesses. Groups push in different directions, competing for dollars and supporters. Rarely do they speak with a common voice. Most troubling is the movement's lack of an overall goal, a clear vision for how people ought to live so as to respect nature adequately. We hear about clean air and clean water; we hear about the need to protect rare species and the wisdom of healthy food. But what about the land itself, the working landscapes where people live? What is our conservation goal for them? Surely they are not sacrifice areas, which we can use however we like so long as we atone by setting aside high-quality preserves elsewhere.

This lack of an overall goal—some vision of good land use—is a serious impediment for defenders of natural areas, and yet it is also an opportunity. Somebody needs to step forward and talk coherently about the well-being of landscapes where people live. Conservation, first and foremost, is about taking good care of nature. So what does that mean in landscapes that people occupy? Surely it entails more than keeping key resources flowing, but what more is it? And who is going to assemble the pieces to yield an answer? With the academy and the conservation movement so fragmented, who is going to talk about conservation as a whole?

One person who did was Leopold. Drawing upon years of

study, he pieced together an overall goal for conservation and then proposed the goal to his colleagues. It was a landscape-scale ideal, grounded in ecology and based on a detailed understanding of how land functioned. Leopold termed his goal "land health." The particular label contained no magic for Leopold. It was the idea beneath the term that counted. The land was best understood as a community of life, Leopold contended, with people included in it. That overall community could be more or less healthy. Land health was thus conservation's rightful goal.[6]

Leopold hoped his goal would catch on, but it did not. Decades later, however, other conservationists began moving close to where Leopold ended up. Today we have technical writings on ecosystem health, ecological integrity, ecosystem services, and other conservation ideas that dwell rather similarly upon nature's processes. So far, though, none of the ideas have gained currency, and few conservation groups are paying much attention to them.

A conservation ideal of this sort, tailored to human-occupied landscapes, is very much needed today. Without one, the conservation movement will continue to drift. We can illustrate this need by giving thought to a land-related statute that Congress enacted in 1999.[7] The statute provided a new charter for the national wildlife refuges of the United States. It represented a rousing victory for wildlife advocates at a time when victories were few. The statute stated clearly that wildlife would come first on the refuges, with other activities fitting in only when consistent. It also proclaimed—and this was even more exciting for ardent wildlife advocates—that refuges would be managed to maintain their "biological integrity, diversity, and environmental health." These words were music to the ears of devoted biodiversity advocates. Finally, the key terms of biodiversity protection had made it into law. Wildlife would come first except in refuges where Congress had expressly said otherwise. It was a happy day.

The statute was indeed a step forward for conservation, but its firm, pro-wildlife language came at a hidden practical cost. The new standard set for managing refuges, phrased in terms of the biological integrity, diversity, and environmental health of the refuges, was a perfectly good goal for the refuges themselves, considered as natural enclaves. But it was emphatically not a sensible goal—not even a plausible goal really—for the larger landscapes of which the refuges were a part. Outside the refuges people had to live and gain food. And they could not do that without violating the integrity-diversity-health standard of the statute, given that Congress had defined these key terms so strictly that nearly any human change to nature amounted to degradation. Refuges could be run that way, but other landscapes could not. People have to change nature to live.

What all of this meant was that wildlife refuges would thereafter be managed according to their own, peculiar land-management standard. Wildlife refuges would promote a special ecological goal that simply could not govern the larger landscape. A refuge would thus have its particular goal; other landowners would have their much different goals; and life would go on. Land boundaries would be respected. Landscapes would remain fragmented into various land-management regimes. There would be no common vocabulary for talking about what all lands had in common.

Some wildlife-refuge managers were quick to see they faced a problem with this strict new management standard. They knew they could manage their refuges to promote integrity, diversity, and environmental health only if they took into account activities occurring on surrounding, nonfederal lands. Ecological processes do not respect property boundaries, and neighboring land uses inevitably affect refuges. Indeed, many refuges are themselves highly fragmented geographically. Animals in refuges regularly wander onto surrounding, nonfederal lands. It was simply not possible to maintain the strict integrity of refuges unless land uses somehow were coordinated at the landscape scale. This was the problem, a big one. And refuge managers could see no good solution to it. Individual managers would simply have to deal on their own with adjacent, clashing land uses—what the statute ominously termed "external threats."

A key piece was missing here. Refuges needed more policy guidance than Congress provided. The missing element was an ecologically based vision that could serve to guide land uses in landscapes as a whole—something like Aldo Leopold's goal of land health. If there were such an overriding goal, and if people supported the goal, then they would all have a common way to talk about their respective land uses.

A shared goal would offer many benefits. It could make clear how all lands within a landscape were ecologically connected and interdependent. Equipped with a shared goal, a refuge manager could then explain how her well-managed refuge helped achieve this larger goal, benefiting everyone in the landscape. A sound goal would also provide useful language for evaluating nearby private land uses. When managers (and statute drafters) talk about surrounding activities as "external threats," they imply a distinct conflict between the goals of refuges on the one hand and the goals of private owners on the other. The situation would appear different if all lands were subject to an overriding ecological baseline. Then the question would not be whether a private land use was harming the refuge but whether it was harming the landscape as a whole. When a landowner undercuts the shared goal, she harms the entire community, not the wildlife refuge alone.

Congress's failing, then, was this. When it wrote the new refuge statute, it remained too tied to the view of wildlife refuges as distinct enclaves. It failed to see that refuges are valuable chiefly because they benefit larger landscapes. Congress should have talked about them that way, and it should have gone on to consider the ecological standards that might apply in whole landscapes, instead of thinking only about the stringent standards applicable to refuges. By prescribing rules for refuges alone and then talking about external threats, Congress largely accepted the fragmentation of landscapes. The resulting 1999 statute does little to reform our ingrained ways of seeing and understanding land.

With these observations on the table, we are in a position to see what it will take to keep natural areas afloat, in their natural landscapes and in the treacherous waters of modern culture.

Natural areas benefit people in many, evolving ways. Managers of natural areas need to stay skilled in talking about these benefits and about the ways refuges never rest in the services they provide. It is not enough, though, to keep a list of human benefits in the top desk drawer, ready to pull out when a natural area requires defending. Managers need to go further. They need to challenge surrounding culture, drawing upon the values embedded in natural areas and pushing people to see land and their place in it anew. They need to resist prevailing values that are alien to what natural areas are about. This means challenging fragmentation in its various forms, promoting ecological thinking about landscapes, and pushing people to reconsider the human-drawn boundaries on the land. Nature's landscapes are seamless wholes. Only people see them as fractured.

Too many American landscapes are in poor condition ecologically. American culture is also in poor condition, in terms of the ways we see nature and situate ourselves in it. The problems are linked, and they are best addressed together. Natural areas and their defenders should be at the forefront of this change. Buildings might come and go, but natural areas must endure. Particular land uses can shift, but the landscapes themselves, as ecological units, should always function productively. According to opinion polls, most Americans believe we bear an ethical duty to protect rare species. Plainly, natural areas are essential in meeting this felt obligation, just as they are essential in ministering to our souls. Natural areas help us live better ethically, and managers need to say so.

Given the industrial, market-based slant of contemporary culture, surely it is a good time to be culturally out of step—proudly, even defiantly so—particularly on matters relating to nature. For those who love land, these are dim times. They resemble in ways the cultural darkness of early medieval Europe. Then, the monasteries that dotted the landscapes kept alive the lamp of learning. They were enclaves of culture in landscapes that had fallen down. One day, perhaps, we will look upon our natural areas as the monasteries of our own age. They keep alive parts of our culture that are essential. When we care for wild places, we harbor a different view of the world, a different set of values and hopes, a different set of dreams.

Notes

1. David Orr, "The Urban-Agrarian Mind," in *The New Agrarianism: Land, Culture, and the Community of Life*, ed. Eric T. Freyfogle (Washington, DC: Island Press, 2001), 98.

2. Aldo Leopold, "Conservation Economics," 1934, in *The River of the Mother of God and Other Essays*, ed. Susan L. Flader and J. Baird Callicott (Madison: University of Wisconsin Press, 1991), 197.

3. *Glisson v. City of Marion*, 720 N.E. 2d 1034 (1999).

4. The roles and limits on science in this area are considered in Eric T. Freyfogle and Julianne Lutz Newton, "Putting Science in Its Place," *Conservation Biology* 16 (August 2002): 863–73.

5. I frame the issue and identify its main elements in Eric T. Freyfogle, "What Is Good Land Use?" in *Why Conservation Is Failing and How It Can Regain Ground* (New Haven, CT: Yale University Press, 2006), 144–77.

6. Aldo Leopold's conservation, including his goal of land health, is ably considered in Julianne Lutz Newton, *Aldo Leopold's Odyssey* (Washington, DC: Island Press, 2006).

7. The statute is considered, and my points here are fleshed out, in Eric T. Freyfogle, "The Wildlife Refuge and the Land Community," *Natural Resources Journal* 44 (2004): 1027–40.

2

A Durable Scale

The conservation community in the United States suffered a loss in April 1948, when sixty-one-year-old Aldo Leopold died fighting a grass fire on a neighbor's farm. It was a loss not just of a lead conservation voice but of a type of conservationist, one who could roll up his sleeves and labor on land yet who understood the broad cultural and economic contexts of land use and conservation. It was the latter skill that made Leopold so valuable, then and now. Even as he mastered scientific details, he developed a rare ability to distance himself from the day-to-day. He could step back from his surrounding culture, assessing its strengths and weaknesses and identifying why people used and misused land as they did. Leopold had little trouble isolating the root problems that conservation faced. The problems were not chiefly scientific, nor did they admit of technological fixes. Land degradation arose because people used the land wrongly. It would be reversed only if they changed their ways.[1]

Over the course of his forty-year career, Aldo Leopold took part in pretty much the full range of conservation work: soil protection, forestry, grazing, waterways, wildlife, agriculture. In dealing with the challenges, he necessarily roamed widely among the sciences. He roamed just as widely in his thoughts about the human condition. Leopold could see how important human behavior was and, hence, how important social and economic factors were. Behavior was closely linked with popular values and widespread assumptions about land and the human place. If better public behavior was going to come about, American culture had to change.

Upon Leopold's death, his work was largely taken over by an array of technicians rather than by conservationists of similar breadth. Experts divided nature study and conservation into many parts and then allocated the specific tasks. This specialization was already advanced during Leopold's day, and he was greatly troubled by it. Scientists were gaining expertise in ever-smaller areas. Few of them were thinking about the landscape in its entirety, with people included. Few could see how the conservation of particular resources—water, grass, timber, wildlife—presented a single challenge rather than multiple challenges. Even fewer recognized how the conservation of land and the conservation of people were intimately linked. The new technicians brought skill to their tasks, but the broader picture was losing focus, and too often the technical parts fit together poorly.

Leopold first gained prominence in 1920, when he promoted the then-novel idea of protecting the nation's remaining wild lands.[2] Wild places had survived because no one had gotten around to altering them. Given the rising economic pressures that were bringing automobiles and new roads, legal protection seemed needed to keep them wild. Leopold urged that it take place. Largely through his efforts, the U.S. Forest Service in 1926 designated the nation's first wilderness, in the Gila National Forest in Arizona. In writings in the 1920s and 1930s, Leopold proceeded to catalog and publicize the reasons that wilderness protection made sense.[3] In 1935 he and six others founded the Wilderness Society, with Leopold writing its culturally charged manifesto. Leopold continued promoting wilderness protection after moving, in midcareer, from the Southwest to the more densely occupied farm region of central Wisconsin. He is remembered, justly, as the founder of wilderness protection and as a particularly able advocate for it.

Leopold's initial defense of wilderness focused on the recreational value of roadless areas for visitors who really wanted to get away. The road-building urge was welling up, and roadless areas were disappearing.[4] To protect big wilderness, Leopold argued, was to preserve for a hardy minority a type of recreational opportunity that could not take place elsewhere. To this recreation rationale Leopold soon added the idea of wilderness as wildlife habitat for big game animals whose numbers were declining because of domesticated livestock and expanding settlements. This second rationale was also human centered. More and more people were buying licenses to hunt, and game populations were falling. More game meant better hunting. Before long, Leopold's thinking about wildlife expanded beyond game species. Wilderness areas were also good refuges for nongame species, particularly animals that had trouble living near people. Even predator protection entered the picture, in a limited way. By the late 1930s, Leopold viewed wilderness areas as essential habitat for the gray wolf, grizzly bear, and other predators that were gone elsewhere. Their presence added to the allure of wilderness adventure while helping to keep prey populations in check.

From the beginning Leopold knew that wilderness areas offered cultural benefits as well as recreational and ecological ones. American culture had taken shape at the wilderness edge, by log-cabin building and the covered wagon. To preserve wilderness was to retain ties to the nation's past and to the landscapes that made us who we are. Wilderness reminded us where we started while perpetuating the "virile and primitive skills in pioneering travel and subsistence."[5] Similarly, wilderness provided places where people could go to shake off the clutter of modern life. It offered a tonic for the anxieties of the age.

Leopold crafted these arguments to persuade diverse audiences to support wilderness reserves. His rationales spoke to people where they were and used the vocabulary of the day—often the highest-and-best-use reasoning made popular by Progressives. Yet, as Leopold continued to reflect on wilderness, he sensed that its main values lay elsewhere. And they were even more subversive of America's dominant culture.

Leopold's further thoughts on wilderness progressed alongside his thoughts on related subjects, particularly on how land functioned and how people ought to inhabit it.[6] Cultures faced real challenges in finding ways of living that could endure. In the Southwest, Leopold could attest from personal study, several civilizations had come and gone. Somehow their ways of living had run afoul of the exigencies of local nature. The challenges were particularly acute in lands that were on a hair trigger ecologically because the margin for error was small. A civilization could last, Leopold decided, only if it respected the land's ecological functioning. And that was not easily done.

The more Leopold learned about land, the more he worried about his own culture. The culture was flawed and required change if people were to thrive, but what new cultural values and practices were appropriate? And how could society get from where it was to where it needed to be? Leopold struggled with the questions. As he did so, wild places became more important in his thinking—big wild places as well as the small scraps that civilization had bypassed. Considerable wisdom lay in these wild places, Leopold sensed. In them, scientists could learn how healthy land functioned. In them, people could identify the kinds of values that were required to tend land well. Enduring civilizations successfully shaped themselves around the land. Presumably American society needed to do the same if it expected to endure. Wilderness, then, could provide a template for cultural change. By studying unaltered land, a people could forge the elements of a lasting civilization. And if their effort failed, they could go back and start again. Or they could if they kept their wild places intact.

By late in life, in short, Leopold had identified further reasons that wilderness preservation made sense. And they were the most important. Wilderness could help bring on the kinds of cultural change that America required to craft an enduring civilization. It was a powerful rationale in Leopold's view, but could it make sense to the American people?

Leopold's training was in forestry. Upon graduating from Yale Forest School in 1909, he joined the ranks of the nation's Forest Service. The service was new, its officers were mostly young, and advancement came fast. After an early professional stumble (soon corrected), Leopold rose to forest supervisor and then to inspector of forests. He quickly gained a knowledge of large landscapes and of the ways people interacted with them. Leopold could see the various roles that forests played in larger, human-dominated landscapes. National forests had been set aside because their ecological effects benefited humans who lived outside the forests as well as those who lived inside them. From his early days, then, Leopold thought in terms of whole landscapes and of how land uses could have wide-ranging effects.[7]

In the mid-1920s Leopold moved to Wisconsin, shifting from a sparsely populated landscape dominated by public lands to a more cultivated landscape fragmented into private parcels. Wilderness preservation was not an issue in southern Wisconsin, and large predators were rare or gone. The local conservation challenges instead were basic ones: keeping soil intact and fertile, maintaining reasonable water flows in quantity and quality, and promoting sound populations of wildlife species that could tolerate human neighbors. Leopold developed an expertise in wildlife management and produced the nation's first scientifically grounded text on game. As a university teacher, Leopold focused on ecology and wildlife. As a citizen and public servant, he addressed a wider array of conservation problems, particularly the ones that nagged Wisconsin farms. He worked actively in organizations and conservation causes. He gave radio talks, taught short courses, participated in study teams, wrote letters and reports. His energy seemed inexhaustible; his range of interests was broad.

As Leopold worked and learned, however, he became more aware of problems within the conservation movement. Conservation efforts tended to focus on individual challenges. Advocates of specific measures too often stumbled over one another. Different efforts competed for money and members. Even worse, they often dispensed conflicting advice to landowners and others. Wildlife was a prime example. Some conservationists called for habitat restoration, others for captive breeding. Some called for restoring native species, others for importing exotics that displayed more appealing characteristics. Some pushed landowners to keep lands open to hunting, whereas others thought that posting land would give landowners a valuable incentive to promote them. Leopold could see how bewildered if not angered a landowner might become. Adding to the troubles was the increasing fragmentation within intellectual thought generally. Leopold lamented that fragmentation in a familiar passage in his classic *Sand County Almanac*: "There are men charged with the duty of examining the construction of the plants, animals, and soils which are the instruments of the great orchestra. These men are called professors. Each selects one instrument and spends his life taking it apart and describing its strings and sounding boards. This process of dismemberment is called research. The place for dismemberment is called a university."[8] When they focused on their narrowly defined goals, even well-trained scientists could fail to see that their work might be misdirected as well as incomplete. A common flaw was that conservation proposals typically dealt with symptoms, not underlying diseases. Leopold bemoaned that

> when a soil loses fertility, we pour on fertilizer, or at best alter its tame flora and fauna, without considering the fact that its wild flora and fauna, which built the soil to begin with, may likewise be important to its maintenance. . . . Many conservation treatments are obviously superficial. Flood-control dams have no relation to the causes of floods. Check dams and terraces do not touch the cause of erosion. Refuges and hatcheries to maintain the supply of game and fish do not explain why the supply fails to maintain itself. . . . The practices we now call conservation are, to a large extent, local alleviations of biotic pain. They are necessary, but they must not be confused with cures.[9]

Leopold's worries about society's predicament strengthened as he learned more about interconnections. Though he contributed only occasionally to the technical literature on ecosystem functioning, he followed it closely.[10] Leopold was particularly influenced by writings on the cycling of nutrients and on soils and the complex ways that soil biology fostered plant growth. His reading, his own research, his practical conservation work, and his interactions with farmers all led him to see that land was best understood as an interrelated community of life—not as tightly knit as an individual organism, but an integrated community nonetheless. And it was a community, he slowly realized, whose complexity exceeded the ability of scientists to unravel. The more he studied, the greater his ignorance seemed to become. The more he explored, the harder it became to grasp why life acted as it did. Most species played communal roles that were essentially inscrutable. That being so, who could decide which species were valuable and to what degree? The Progressive Era conservation movement was premised on the notion that humans could manage lands to maximize the long-term yield of useful life forms. But what happened when the line between the useful and the useless blurred or disappeared? Some species obviously had direct value for people. As for the rest, who was to say? Leopold for one believed he could not. What he did know was that the whole of nature was important and that this natural whole could be more or less healthy in its functioning. It could be more or less fertile and productive, more or less able to withstand stresses and continue its evolutionary and ecological ways.

Leopold's thoughts about utility and ecology showed up in his writings on wildlife.[11] His early writings spoke of managing land to yield crops of animals to shoot and trap. Soon he saw that nongame animals were also important and that the management of land to maximize desired species could easily harm the larger land community. Ironically, Leopold, the first professor of game management, began to doubt the entire idea of managing wild animals as such. Excessive numbers of a given species could harm the land, and so could distorted mixes of species. Perhaps some species were not necessary, and perhaps populations of other species could be increased without causing damage. But who knew enough to make the decisions?

The 1930s was a time of awakening for Leopold. Much of it came during a lengthy trip to the Sierra Madre range in the Mexican state of Chihuahua. There he saw land that, to his trained eye, appeared unaltered by human hands. He was startled to see how different it was from the climatically similar lands he had known nearby in the southwestern United States. Its waters were clearer, its river banks more lush, the overall landscape more fertile. As he would later write in a brief memoir, "All my life I had seen only sick land, whereas here was a biota still in perfect aboriginal health. The term 'unspoiled wilderness' took on a new meaning."[12]

During the mid- and late 1930s, things finally came together in Leopold's mind. Land was best understood as a community of life, a community that could be more or less fertile, productive, and thus healthy over the long run. The land's health, in turn, depended in complex, unfathomable ways on the various life forms that composed it and that interacted to keep it going. Healthy functioning, he knew, was intimately tied to the land's ability to keep soil fertile, its ability to recycle nutrients over and over so that death led to decay and gave rise to new life. But beyond that, who could say what kept land healthy? Leopold presumed that most, if not nearly all, species in a given landscape played roles. But what were those roles, and which species were more important? Which if any might be eliminated or replaced without diminishing the whole? Leopold the scientist was reluctant to answer.

From this point on, Leopold talked regularly about land sickness and land pathology. Again and again he ticked off for audiences what he viewed as the key symptoms of illness. A typical summary appeared in his important essay "The Land-Health Concept and Conservation": "The symptoms of disorganization, or land sickness, are well known. They include abnormal erosion, abnormal intensity of floods, decline of yields in crops and forests, decline of carrying capacity in pastures and ranges, outbreak of some species as pests and the disappearance of others without visible cause, a general tendency toward the shortening of species lists and of food chains, and world-wide dominance of plant and animal weeds. Without hardly a single exception, these phenomena of disorganization are only superficially understood."[13]

Around 1939 Leopold made an important linguistic shift, changing from talk about land sickness to affirmative talk about land health. Once this idea came to him, it pushed itself into the center of his thoughts on conservation, land, and the human predicament. Yes, he seemed to say to himself, land was a community of life, and humans were part of that community. Humans needed to live in ways that fostered land health; it was as simple and yet as naggingly complex as that. Land sickness arose because humans were not fostering land health. Conservation, in turn, was a matter of restoring health. Conservation in all of its aspects and complexities had to do with bringing human ways into alignment with the land community.

Land health was probably the most important conservation idea that Aldo Leopold produced. If understood better and more widely, it could stand as the single most important conservation idea that America has yet produced. Once Leopold latched on to land health, all else that he did and nearly all that he thought were subordinate to it.

Although Leopold felt secure about his brief summaries of land health, his confidence flagged when he tried to be precise about what it took for land to function well. Human knowledge was simply too elementary. "As a matter of fact," he wrote in 1944, "the land mechanism is too complex to be understood, and probably always will be. We are forced to make the best guess we can from circumstantial evidence."[14] He wrote in 1942, "The best we can do [by way of understanding and promoting land health] is to recognize and cultivate the general conditions which seem conducive to it."[15] Yet Leopold did not accept ignorance as an excuse for inaction, particularly by experts. Conservation problems were grave, and prompt action was needed. In a 1946 essay left in pencil draft at his death, he issued a plea for his fellow ecologists to join him in offering their "best guesses" as to what it took for land to regain health. Their knowledge was plainly imperfect, but they knew more than untrained observers. Their best guesses were "likely to contribute something to social wisdom which would otherwise be lacking." "I have no illusion," he wrote in the same paper, "that the thousands of ecological questions raised by modern land-use can all be assessed by ecologists. What I mean by 'prediction' is a shrewd guess on just one basic question: What are the probable conditions requisite for the perpetuation of the biotic self-renewal or land-health? This would define a goal for conservationists to strive toward. They now have no basic goal bracketing all component parts. Each group has its own goal, and it is common knowledge that these conflict and nullify each other to a large degree."[16]

Once land health came to Leopold, he used it extensively in his writings, expressly and in variant forms. A characteristic use occurred in a 1941 essay, "Planning for Wildlife," which also displayed Leopold's careful attention to conservation means and ends. By then, Leopold was the nation's expert on wildlife management.

He had written widely on the whys and hows of managing animals. But land health as a goal simplified land issues considerably.

> No one can write a plan for accomplishing something until the reasons for desiring to accomplish it are defined. The reasons for restoring wildlife are two:
>
> 1. It adds to the satisfactions of living.
>
> 2. Wild plants and animals are parts of the land-mechanism, and cannot safely be dispensed with.[17]

Protecting wildlife, in short, was a key to promoting land health as well as a means of making life more satisfying.

Leopold is best remembered today for his unusual book *A Sand County Almanac and Sketches Here and There*, which appeared in 1949, a year after his death. The book gained prominence in the 1960s, when it became a bible of the environmental movement. Leopold died before the manuscript underwent editing by its publisher, and he was not around to give consent. A critical editorial decision was to rearrange the order of the four essays contained in the book's final section, "The Upshot."[18] As published, the ultimate essay is the now-famous piece "The Land Ethic," in which Leopold urged readers to cease thinking about land use solely in economic terms and to consider it in ethical terms as well. Leopold's land ethic is more often quoted than understood. Few readers recognize that his briefly presented ethic builds upon a complex understanding of how land functions. It also builds upon, and expressly incorporates, Leopold's vision of land health. The land ethic translates land health into an ethical aspiration for all landowners as they use their lands. It is hardly possible to grasp Leopold's ethic apart from land health and apart from his understanding of how land works.

In Leopold's plan the book would have ended with a relatively brief essay, "Wilderness," his final writing on the subject. The essay covered ground that Leopold readers would have found familiar. Wilderness preserved a special, primitive form of recreation linked to America's pioneer history, Leopold reminded readers. Wilderness was also valuable in protecting wild species. To these points Leopold added a newer argument. Wild areas had value to ecologists as places to study nature's components and processes. So important were these wild places that the Ecological Society of America, upon its formation during World War I, immediately set up a committee to identify and protect the best places.[19] Wilderness was invaluable to science, a point Leopold mentioned in his Wilderness Society manifesto of 1935.[20] By 1941, Leopold's perspective on this issue had sharpened. To understand land health, much less to promote it, would require a far greater understanding of how land worked ecologically. Research was necessary, vast amounts of it. Wilderness areas could serve as research centers—as laboratories, as he put it—for this long-term work. Wilderness provided an example of land at its healthiest. By studying it, one could see how nature worked free of significant human alteration. That knowledge in turn could help scientists learn how humans had altered other places and find ways to restore health. Leopold did not believe that only wilderness areas were healthy. Human alteration did not always bring decline; indeed, human change sometimes made land better. In upholding wilderness as a laboratory, Leopold was making quite a different point. Wilderness provided a place to learn how nature functioned. Lessons learned there could help guide land-use practices elsewhere. By knowing how wilderness areas protected soil and built fertility, using an array of plant and animal species, we could learn how to maintain fertility in the places where we lived and worked.

Leopold covered these points in his final wilderness essay in the *Almanac*. Then, having made the case for wilderness as a land laboratory (and a place for recreation and wildlife), he ended his essay with something new. He took up the matter of culture, and of the role of wilderness in shaping it. "All history," Leopold observed, "consists of successive excursions from a single starting-point." That starting point was wilderness—the raw land itself, functioning in ways that kept it productive. Wilderness had provided "the raw material" out of which humans had hammered "the artifact called civilization." Sometimes, Leopold implied, that artifact proved durable. Sometimes it did not. When humans stumbled, it was time to start over. It was to that starting point—to the land itself—that humans returned "again and again," launching out on "yet another search for a durable scale of values."[21]

Leopold did not elaborate on his observation, save to repeat a comment he offered in the essay's beginning, that wilderness gave "definition and meaning to the human enterprise." Leopold intended his *Almanac* for a general audience, landowners above all. He could lead the typical reader only so far. Indeed, the entire final part of the *Almanac*, including his essays on wilderness and the land ethic, were intended (as he put it in the foreword) only for "the very sympathetic reader" who wished "to wrestle with the philosophic questions."[22] Even with such readers, Leopold felt constraint. Perhaps only a few of them would understand his embedded message, but they would have to suffice.

Leopold emphasized his comments on wilderness and culture by placing them on his manuscript's final page. Thus he concluded his manuscript not by providing a final answer but by taking his reader back to wilderness and encouraging her to imagine a new start. What if we did go back to the land and begin again? What if we set out to forge a new set of cultural values, ones better able to endure? What would those values be? The questions were not hypothetical. American society needed to do just that, in his view. It needed to go back to the land and try again to construct a sound agriculture.

Leopold lamented how little ecologists knew about land, and thus how much work was needed to give content to a sound conservation goal. The science of land health was in its infancy and needed to mature quickly. But the bigger challenge facing America was the cultural one. People did not perceive land in the integrated way Leopold had come to see it. They did not realize that all life formed a community, humans included. They overlooked ecological interconnections; they undervalued nature's many small parts; they failed to appreciate how their well-being turned on maintenance of basic ecosystem processes—and, so far as Leopold could tell, on the retention of nearly all of nature's living parts. People acted as autonomous individuals, landowners above all. They were too much out for themselves, rarely asking what it meant to live as a responsible community member. Life was all about rights, with little talk of duties.

This realization was hardly good news for Leopold. Indeed, though he kept his chin up, it is easy to imagine that it hit him hard. For people to live in ways that sustained land, they had to undergo a serious shift in thought and sentiment. They needed an ecological awakening. They needed to broaden their senses of value. Perhaps most of all, they needed to become far more humble, to cleanse themselves of their "Abrahamic concept of land"—their belief that nature existed for humans to alter as they saw fit. Leopold expressed the point in the *Almanac*'s foreword: "We abuse land because we regard it as a commodity belonging to us. When we see land as a community to which we belong, we may begin to see it with love and respect. . . . That land is a community is the basic concept of ecology, but that land is to be loved and respected is an extension of ethics."[23]

Leopold would repeatedly play upon this theme of cultural transformation, pushing his readers onward in writing after writing, speech after speech. "It is a century now," he would write, "since Darwin gave us the first glimpse of the origin of species. We know now what was unknown to all the preceding caravan of generations: that men are only fellow-voyagers with other creatures in the odyssey of evolution. This new knowledge should have given us, by this time, a sense of kinship with fellow-creatures; a wish to live and let live; a sense of wonder over the magnitude and duration of the biotic enterprise."[24] It should have given us these things, he said, but it had not. Indeed, progress was hardly perceptible. A major cultural shift was required soon, given the land's sagging health. But how could it come about? Doctoring the land back to health was labor enough; inducing cultural change was even more daunting.

These were the main points in Leopold's mind when he sat down to compile his *Almanac*. In his book and with it, he hoped to promote a cultural rebirth. Indeed, it is revealing to view Leopold's book as a circular creation—a wreath, as poets term it—in which the ending point links to the beginning. In Leopold's manuscript, the ending point was a call for a new beginning—starting over with land in search of a new scale of values. If, upon reaching the end, we return to the book's beginning, we find Leopold ready to lead us on just such a search. The *Almanac* opens with an alluring tale of skunk tracks in the snow, "January Thaw."[25] Leopold invites us to journey with him, following the skunk and imagining its motives for rambling. Like the skunk in winter, we Americans have been asleep to nature. We too need to awaken to the land, to wander out and see it afresh.

"January Thaw" is hardly 650 words, yet it is packed with raw material for a new start. The skunk tracks mark an early event in the land's "cycle of beginnings and ceasings," what "we call a year" (given our tendency to fragment time) but which nature views as a continuous expanse. Taking us by the hand, Leopold displays his overflowing curiosity about nature. His curiosity, he hopes, will prove contagious, as will his love of all things "natural, wild, and free." A meadow mouse darts across our path, prompting Leopold to compare the world of mice and the world of men. Much like a person, the industrious mouse has created beneath the snow a "maze of secret tunnels." But then comes the thawing sun, which "has mocked the basic premises of the microtine economic system." And so too our own economic system, so often constructed with little regard for nature's forces and ultimately just as vulnerable. A rough-legged hawk swoops upon the mouse, "who could not wait until night to inspect the damage to his well-ordered world," and thus we learn a lesson in patience and respect. For the mouse, "snow means freedom from want and fear." For the hawk, it is the reverse—it is the thaw that brings the same freedom. And thus each species evaluates nature from its own perspective, not appreciating its bias. Is the human view any broader? Indeed, do we differ fundamentally from other species in our dependence on nature and limited knowledge of it? If we do not, might we turn to these species as elders to learn the secrets of survival? Ultimately, tour guide Leopold returns home with questions about the skunk's behavior and few answers, reiterating how mysterious nature remains despite our efforts. The natural world is full of secrets, and to unravel one secret is to gain entry to another. What better way to spend time than engage with them!

"January Thaw" begins what is essentially an extended, discursive, engaging commentary on where our culture stands and how it might change. In essay after essay, Leopold expands upon his points. Repeatedly he draws parallels between humans and other life forms. Again and again he illustrates how little we know and how enjoyable nature study can be. He invites us to share his preference for native species and his joy when finding patches where native communities survive. He stretches our time horizons far back and far forward while offering lessons on evolution, fitness, and ecological change. Human-drawn boundaries on land appear artificial from nature's perspective. How presumptuous of us to think that we are the land's "owners" or even its sole tenants. How blind we so often are to nature's ways, even to the fundamental processes upon which our economies depend. Leopold shakes his head at the Phi Beta Kappa graduate who is blind to geese migrations: "Is education possibly a process of trading awareness for things of lesser worth?"[26] Leopold is attentive to the geese and carefully notes their preference to feed on corn stubble that occupies former prairie rather than woodland. Might the geese know something that we do not, he muses—perhaps that prairie soil yields corn of higher nutritional value? The bur oak draws praise as an evolutionary achievement; as the only large tree that can withstand prairie fires, it symbolizes successful adaptation to harsh surroundings. Just so there is *Draba*, the "smallest flower that blows," held up for thanks because it is "a small creature that does a small job quickly and well."[27] Leopold laments the loss of the passenger pigeon yet takes heart in our collective sadness about its parting: "For one species to mourn the death of another is a new thing under the sun."[28] A new thing, and a better one. Meanwhile, traveling by bus through Illinois, Leopold is grieved by the ignorance of corn farmers about the native plants that give rise to the plains' fertility. Weeds, all of them, the modern corn man implies, "his fertilizer bill projecting from his shirt pocket." Though the farm looks solvent, the "old oaks in the woodlot are without issue. There are no hedges, brush patches, fencerows, or other signs of shiftless husbandry. The cornfield has fat steers, but probably no quail." And on it goes, nature's "useless" parts pushed aside to promote a small number of favored species, as if we know what it takes for land to work well. "In the creek-bottom pasture, flood trash is lodged high in the bushes. The creek banks are raw; chunks of Illinois have sloughed off and moved seaward. Patches of giant ragweed mark where freshets have thrown down the silt they could not carry. Just who is solvent? For how long?"[29]

Although land health gains express mention only in the final essays of the *Almanac*, the idea looms above every essay, and Leopold's cultural ideals are intended to promote it. The ideals go further than that, however, for his conservation thought extended beyond ecology and fertility. Humans needed more than smoothly functioning ecosystems to enjoy life fully. The land could be a source of great beauty and in other ways add joy to living. Leopold exalted these joys, in his life and his writing. Also important to Leopold was the whole question of what he termed the "biotic right" of other species and communities to exist. Leopold put this idea at the edge of his work and spoke of it only occasionally. One senses from his written legacy, however, that he strongly felt a moral duty to preserve other species without regard to their ecological value. Perhaps he rarely spoke of biotic right or intrinsic value because he thought that land health was goal enough for any one generation. More likely he kept quiet because the quest for land health was best pursued by doing everything possible to preserve all species. All species might not be needed, and sometimes exotics could successfully replace native species. But because no one really knew which species could safely disappear, the wise approach was to save all the parts.

With these observations in hand, we can assemble the reasons that Leopold, late in life, viewed wilderness as so essential culturally—indeed, as the starting point for a new land-based culture.

Wilderness preservation—putting land off-limits to intensive use—required an act of restraint. The labor of preserving and protecting wild places gave people chances to gain and to exercise humility. When a people labeled a place wilderness, they showed the world they could contain their appetites.

Wilderness was also valuable because it provided lessons on

how land works and what it takes for land to retain health. Many lessons were ecological, but cultural values came along with them. The land provided clues on what humans should value and how they should live. Good land use required careful attention to the peculiarities of a given place. That attentiveness could arise only within a person who loved the land and felt attached to its many inhabitants. These conclusions and related ones, Leopold believed, formed an integrated set of values and understandings. Together they gave structure to a land-based culture tailored to place and thus likely to last.

As it gave rise to new values, wilderness also provided a means of promoting them. The humble act of protecting wilderness helped. The study of nature also helped, particularly when study blended with recreation to become a pastime. Among the many types of outdoor recreation, nature study stood atop Leopold's list as most valuable. Hunting involved killing and was too focused on a few species. Photography dispensed with killing but was otherwise narrow in scope; it too was a form of taking trophies. Merely enjoying scenery and fresh air did little to draw attention to nature's intricacies. Only nature study did that. So much remained unknown about nature that amateur naturalists could still contribute to science. Whether or not they did, study promoted ecological understanding and brought an attentiveness to nature—key elements in a new culture. It was hard to protect a rare plant that one never noticed. It was hard to object to ecological derangements when nature's processes were unknown.

Next to nature study was wilderness travel, taken in the right spirit and under primitive conditions. To live simply in the wilds, avoiding too many gadgets, instilled a sense of connection to nature. Leopold saw value "in any experience that reminds us of our dependency on the soil-plant-animal-man food chain, and of the fundamental organization of the biota."[30] Wilderness travel could do that. Primitive travel and camping meant immersion in nature. It could instill senses of awe and love. This was literally a type of re-creation, of the heart and the mind.

From these protection rationales, we can see what types of lands Leopold viewed as wilderness, in terms of size and ecological condition. Some of his rationales required wild areas vast in size—hundreds of thousands of acres, even larger. Grizzly bears needed room to roam; lengthy wilderness camping trips called for similar expanses. Other rationales, though, required far less space. Scientific research on land health could occur on a handful of unaltered acres. Indeed, key grassland studies and other research had taken place on very small tracts. Cultural awakenings, ecological understandings, senses of nature's complexity and beauty—these too could arise in small places. Predictably, Leopold never set a minimum-size requirement for protection. Nor did he prescribe any necessary degree of freedom from human alteration, though he clearly preferred as little change as possible. Wilderness was a matter of degree, "a relative condition," rather than a matter of all or nothing. Wilderness existed "in all degrees, from the little accidental wild spot at the head of a ravine in the Corn Belt woodlot to vast expanses of virgin country."[31] Preserve the best that is available, Leopold argued, all around the country and on all types of land. Study these places; visit them; draw inspiration and be transformed.

In the dark hours of World War II, Leopold took time to illustrate the ways wilderness could bring transformation. He set his writing on the Flambeau River in northwestern Wisconsin, long a wilderness haven but by 1943 "on its last legs" because of "cottages, resorts, and highway bridges [that] were chopping up the wild stretches into shorter and shorter segments."[32] Evening was coming on as Leopold camped beside the Flambeau. Along came two youths in a canoe, two days out in the wilds. The youths were soon to enter the army and were enjoying a taste of freedom before beginning regimented life. Scantily provisioned, the young men were dependent upon themselves and nature. "No servant brought them meals: they got their meat out of the river, or went without," Leopold observed. "No traffic cop whistled them off the hidden rock in the next rapids. No friendly roof kept them dry when they misguessed whether or not to pitch the tent. No guide showed them which camping spots offered a nightlong breeze, and which a nightlong misery of mosquitoes; which firewood made clean coals and which only smoke."

The freedom that the youths enjoyed was a particularly instructive one, for them and for society. This was not simply the freedom to come and go without oversight. It was the freedom to make mistakes. Immersed in the wilderness, the youths were getting "their first taste of those rewards and penalties for wise and foolish acts which every woodsman faces daily, but against which civilization has built a thousand buffers." "Perhaps every youth," Leopold observed drily, "needs an occasional wilderness trip, in order to learn the meaning of this particular freedom."

Youths needed wilderness trips, and so did we all. Most of us had lost touch with nature and its ways. Buffered by civilization, we were losing our sense of dependence. Nature took orders from us, not the reverse, or so it now seemed. The trends were troubling, even ominous. The "shallow-minded modern" was losing "his rootage in the land."[33] He needed to regain it, and there was no better way than to return to the wilds.

Notes

1. Leopold's science and evolving ideas about land conservation are considered most fully in Julianne Lutz Newton, *Aldo Leopold's Odyssey* (Washington, DC: Island Press, 2006). His life is ably recounted in Curt Meine, *Aldo Leopold: Life and Work* (Madison: University of Wisconsin Press, 1988).

2. See Daniel J. Philippon, *Conserving Words: How American Nature Writers Shaped the Environmental Movement* (Athens: University of Georgia Press, 2004), 159–218; Roderick Frazier Nash, *Wilderness and the American Mind*, 4th ed. (New Haven, CT: Yale University Press, 2001), 182–99.

3. Leopold's chief writings on wilderness are listed in Aldo Leopold, *The River of the Mother of God and Other Essays*, ed. Susan L. Flader and J. Baird Callicott (Madison: University of Wisconsin Press, 1991), 349–70. Along with these pieces—several published for the first time in this edited volume—Leopold left behind a handful of never-published manuscripts dealing with wilderness, all located in the Leopold archives at the University of Wisconsin–Madison. The unpublished writings include "A Wilderness Area Program" (ca. 1922), "Scarcity Values in Conservation" (ca. 1924–25), "Wilderness Conservation" (1926; an ad-

dress to the National Conference on Outdoor Recreation), "The Conservation of Wilderness Recreation Areas" (1927), and "What Shrinks Wilderness" (1940). Comments on wilderness also appear in several of Leopold's early forest inspection reports from the Southwest.

4. The effect of the good roads movement on the wilderness impulse is considered in Paul S. Sutter, *Driven Wild: How the Fight against Automobiles Launched the Modern Wilderness Movement* (Seattle: University of Washington Press, 2002).

5. Aldo Leopold, *A Sand County Almanac and Sketches Here and There* (New York: Oxford University Press, 1949), 192.

6. The intertwining character of Leopold's conservation ideas forms the organizing thesis of Newton, *Aldo Leopold's Odyssey*.

7. For the facts in this paragraph and those that follow, I have drawn from the important studies by Newton (*Aldo Leopold's Odyssey*) and Meine (*Aldo Leopold*).

8. Leopold, *Sand County Almanac*, 153.

9. Ibid., 195–96.

10. Leopold's engagement with ecology (including his service as president of the Ecological Society of America) is set forth in Newton, *Aldo Leopold's Odyssey*.

11. An important early study of Leopold's ecological ideas about wildlife is Susan L. Flader, *Thinking Like a Mountain: Aldo Leopold and the Evolution of an Ecological Attitude toward Deer, Wolves, and Forests* (Columbia: University of Missouri Press, 1974). His ideas about the cultural values of wildlife, about hunting, and about how wildlife study might stimulate cultural change are best told in Newton, *Aldo Leopold's Odyssey*. Further insights are offered in Curt Meine, "Emergence of an Idea," in *Correction Lines: Essays on Land, Leopold, and Conservation* (Washington, DC: Island Press, 2004), 117–31.

12. The quote comes from a draft foreword to *A Sand County Almanac* that Leopold wrote but then set aside. It was published years later in J. Baird Callicott, ed., *Companion to* A Sand County Almanac*: Interpretive and Critical Essays* (Madison: University of Wisconsin Press, 1987), 281–99 (the quote is on 285–86).

13. Aldo Leopold, "The Land-Health Concept and Conservation," 1946, in *For the Health of the Land: Previously Unpublished Essays and Other Writings*, ed. J. Baird Callicott and Eric T. Freyfogle (Washington, DC: Island Press, 1999), 219.

14. Aldo Leopold, "Conservation: In Whole or in Part?" 1944, in *River of the Mother of God*, 315.

15. Aldo Leopold, "Biotic Land-Use," 1942, in *For the Health of the Land*, 203.

16. Leopold, "Land-Health Concept and Conservation," 220.

17. Aldo Leopold, "Planning for Wildlife," 1941, in *For the Health of the Land*, 193–94.

18. Meine, *Aldo Leopold*, 524.

19. Robert Croker, *Pioneer Ecologist: The Life and Work of Victor Ernest Shelford, 1877–1968* (Washington, DC: Smithsonian Institution Press, 1991), 120–44.

20. Aldo Leopold, "Why the Wilderness Society?" *Living Wilderness*, September 1935, 6.

21. Aldo Leopold, "Wilderness," in *Sand County Almanac*, 188, 200.

22. Leopold, *Sand County Almanac*, viii.

23. Ibid.

24. Ibid., 109.

25. Ibid., 3–5.

26. Ibid., 18.

27. Ibid., 26.

28. Ibid., 110.

29. Ibid., 117–19.

30. Ibid., 178.

31. Aldo Leopold, "Wilderness as a Form of Land Use," 1925, in *River of the Mother of God*, 135.

32. Leopold, *Sand County Almanac*, 112–13.

33. Ibid., 200.

3

THE EDUCATION OF ADA

It is early one morning, August 1864, in the mountains of western North Carolina. Ada Monroe has risen and sits on her house porch. The life she has known has wound down and come to a halt. Kinless and nearly friendless, alone and immobile, she has no idea what to do. The solace she gains from books and art is not enough to sustain her. Though Ada does not realize it, the next twenty-four hours will set her on a new course. She will begin life again, in a world badly torn by war. Her central possession is a three-hundred-acre hilly farm inherited from her recently deceased father. Farming is alien to Ada, and she knows little about practical living. But the farm provides a place to begin. And so, slowly, she moves, awakening to the land and shaping a new life.

Ada Monroe's journey is recounted in Charles Frazier's agrarian novel, *Cold Mountain*, published in 1997 to public and critical acclaim. Frazier has a sensitive eye for people, culture, and land. In his narrative he exposes and probes the challenges we face in living with nature and one another. Frazier's work raises a possibility similar to the one posed by Aldo Leopold in his final wilderness essay. When a culture stumbles and falls, is it possible to turn to land to start again? Can we begin with land and craft ways of living that are better suited to place and more likely to endure? Leopold turned to wilderness to begin that search. Frazier proposes beginning nearby, on a subsistence farm carved from the woods. Can such a farm, Frazier implicitly asks, provide a new cultural beginning in a time of war? And can such a land-based culture really succeed, given our passions and weaknesses as people? Like Leopold, Frazier offers only fragments of answers. But they are sturdy fragments, valuable in thinking about the course ahead.

Ada's twenty-four-hour descent and resurrection begin, appropriately, in hunger. From her porch she surveys the nearby garden, overrun by weeds that she can neither identify nor fight. Remembering her hen's practice of hiding in nearby boxwoods, she folds her skirt tightly and works her way on hands and knees into them, searching for eggs. Inside the shrubs she finds "a hollow place . . . like a tiny room." There she pauses to consider her life. She has returned to the womb and is ready for rebirth.

Raised in the high culture of Charleston, South Carolina, by her minister father, Ada spent timing learning French and Latin, the piano, and landscape sketching. She is unusually well read, and her many opinions are colored by her late father's transcendentalism. But Ada is "frighteningly ill-prepared" in the skills of daily living, and she knows it. Resting in her small place, she wonders whether her upbringing could have been any less practical. She admits—echoing Thoreau and his bean field—that she could hardly weed a row of bean plants without mutilating them. Her urban culture has become largely useless to her, and by steps she puts it behind: her dresses get dirtier, her hair hangs loose, her piano grows quiet. Finally, pushed out of the shrubs by nature (an aggressive rooster) rather than by choice, Ada returns to the house. There she gains her first inkling of direction by glancing up from a grim novel and onto the looming shape of Cold Mountain. There is a solidity to it, a comfort. After a failed effort to bake bread, she leaves the house and enters the weedy, little-used lane. Her new life has started. Her first act is to approach an unfamiliar plant. When a blossom snaps apart at her touch, she gives the plant a name—snapweed. She has begun to take charge of Eden.

Ada's path leads her to a nearby farmstead, the home of Esco and Sally Swanger. They are successful farmers, so self-sufficient that they possess a water well despite an abundance of local streams. Their discipline is apparent in a clean-swept yard kept free of ornamental bushes and flowerbeds. For Ada and for us, the Swangers offer a vision of relating to land and neighbors. Frazier invites us to assess it. In mid-afternoon Ada departs the Swanger farm, following an old footpath up the mountainside through second-growth timber and on to a ridge of ancient trees. There, in this wilder place, Ada halts on a rock outcrop to survey her farm. Unless work is done on it, she realizes, weeds will take over. Carefully she traces its outlines in her mind. The farm, we see, has entered her and is taking root. Descending from the crest, she reaches the stone wall at the edge of her upper pasture. It is the place, fittingly, where the wild and the pastoral, nature and culture, come together. Sleep takes over. During the night Ada awakens and then stays to watch the rising sun. The land around her, she recognizes, is all that she can now count on. A future life slowly begins to appear, "a satisfactory life of common things . . . though she could in no way picture even its starkest outlines."

As morning comes Ada finds herself back at her house, wondering how she might begin this new life. Walking up her lane, fortuitously, is another young woman, Ruby—raw gem of the earth. Ruby's life path has been the opposite of Ada's, all nature and little culture. A homeless child of the wilds, raised by a dissolute, mountain-roaming father, Ruby is bound to the land in its wild and domesticated forms. Yet she can neither read nor write, spurns society, and is disconnected from life beyond the hills. Where Ada has land and no skills, Ruby has skills and no land. And so they forge a bond, a marriage of nature and culture on the farm. Their resulting partnership forms one of *Cold Mountain*'s two narrative spines. Aided by Ruby's knowledge, Ada roots herself to her home cove and particular farm. Her attentiveness to nature rises up as her interest in the distant world dims. On her side, Ruby slowly realizes that people can benefit by working together. In time the women dream about the farm and about the kind of life they might make on it. As readers we join in their visions.

Interwoven with Ada's and Ruby's efforts to dwell in Black Cove is the novel's other story line. A wounded Confederate veteran has escaped from a military hospital and is making his way home to Cold Mountain. Inman is larger than life, just like Frazier's other characters. His wounds to neck and hip, though serious, are modest compared with his inner damage. He is badly fractured in spirit—haunted by memories of battlefield slaughter and repulsed by man's darker side. Inman and Ada have a vague understanding of marriage and have written occasionally during his absence. As Inman works his way west, hiding from home guards who brutalize runaway soldiers, he dreams of a new life with Ada on a mountain farm. Perhaps there, cut off from the downward-sliding culture of the age, he can regain wholeness. Inman's journey contains echoes of many literary works—*The Odyssey*, *The Canterbury Tales*, *The Divine Comedy*, *The Adventures of Huckleberry Finn*. It invites comparison, too, with biblical narratives. By his example Inman leads his war-ravaged people on an exodus toward a land of new hope, to a land that he will glimpse but, like Moses, never inhabit. And just as victory lies in sight, Inman will turn the other cheek to a home guard youth—his ensuing death a sacrifice followed by scenes of redemption.

Frazier's richly imagined world is distant from our own, but his characters, beneath the surface, are easy to recognize. And we have no trouble recognizing the challenges they face, for we too must find ways of living on land that can endure. Unruly nature sets the stage, for them and us. More challenging are the troubles that arise within and among people, particularly the tendencies toward wanderlust, waywardness, and violence that Frazier displays vividly. For a responsible, caring people to live well on land is one ambition. For such a people to find places and ways of containing the misfits among them—and still live right on land—is far more challenging. What might lie ahead for Ada Monroe and for the people gathering with her in the shadows of a timeless blue mountain? Can they successfully begin with land—can we begin with land—and construct a culture that can thrive?

Frazier uses a vivid narrative scene to begin his illuminating meditation on nature and culture. Inman lies in a military hospital, recovering from a deep gash to his neck. The ward is dark and his spirit is broken. He has "burned up the last of his own candles," literally and figuratively. A veteran of many fights, "he had seen the metal face of the age" and come away badly hurt. In the cot next to Inman lies another wounded man—Balis, a university-trained Greek scholar who spends his hours translating ancient texts as gangrene in his leg requires amputation after amputation.

Inman's battlefield injury was so severe that fellow soldiers left him to die. When he did not, he was removed to a hospital tent where doctors again viewed his case as hopeless. And yet he has survived. Human knowledge, we realize—the human power to heal—has limits. If Inman is to live, nature must do the work. As the healing takes place, nature expels from Inman's neck various emblems of human culture—a button, a piece of wool from his collar, and a chunk of gray metal. In this tiny setting at least, nature is regaining control.

Inman assists in his own healing and sustains his will to live by reaching out to nature. Hour by hour he observes the nature outside his window, carefully noting every change in it. That nearby nature, in turn, reminds him of his homeland in the mountains, which supplies further strength. Then there is the book that he reads and rereads. It is a torn volume of William Bartram's *Travels*, a classic, vivid narrative of eighteenth-century nature travels in the South. The nature book never fails to ease Inman's thoughts. It settles his mind, particularly the passages that recount Bartram's wanderings near Inman's mountain home. Inman's volume lacks one cover, so he removes the other and rolls the book into a scroll. Bartram has become his holy text, a touchstone for a new, land-based order. Inman keeps the scroll close and refers to it in times of trial. Inman's deity, it appears, resides on earth and in its living creatures, not in the sky.

Frazier's opening scene casts culture in an unfavorable light. Nature is the healer, culture the destroyer. To commit to nature is to gain strength; to wallow in culture—even in Greek classics—is to lose vitality. Thus we see Inman using his shared window to connect to healing nature; in contrast, Balis the Greek scholar, soon to die, uses the window only as a place to spit. Frazier's comparison here appears harsh: Balis's spitting typifies the ways most people disdain nature. Yet Frazier's initial nature-culture clash quickly and appropriately becomes complex. Nature heals Frazier but by no means everyone else; it is unreliable and unpredictable. Walking on the hospital grounds, Inman confronts a blind man whom he has watched boiling peanuts and selling newspapers. He assumes the man's blindness comes from human malfeasance and is surprised to hear the man was born to it. Nature too can be cruel, Inman sees.

His strength returning, Inman walks to town to buy clothes and practical supplies. He tosses his old hat aside, announcing his hope to become a new man. But reclothing the outer man is far easier than rebuilding the inner one. That night Inman escapes with his few belongings. His destination is Cold Mountain, which has "soared in his mind as a place where all his scattered forces might gather." As place, symbol, and dream, Cold Mountain anchors Inman's life. "He thought of getting home and building him a cabin on Cold Mountain so high that no soul but the nighthawks passing through the clouds in autumn could hear his sad cry. . . . And if Ada would go with him, there might be the hope . . . that in time his despair might be honed off to a point so fine and thin that it would be nearly the same as vanishing."

Even with such dreams, though, Inman's despair runs deep. He cannot free his mind from the battlefield slaughter at Fredericksburg and from the quiet pleasure Southern generals seemed to take in it. Inman's feelings grow even darker as he crosses the eroded and degraded North Carolina tidewater. The once-tall trees have largely given way to scrubby pines and poison ivy. A wide river that he encounters is foul and sorry, and he is repulsed by the "monstrous flabby fish" in it, their "meat as slack as fatback." These slow-moving bottom feeders compare badly with the trout in the cold streams of Cold Mountain, "bright and firm as shaving from a bar of silver."

Inman's troubles are compounded by the human evilness he encounters. Three men try to steal his few belongings, one a smithy —a worker in metal—who awkwardly attacks Inman with a scythe and then with a pistol. With his agrarian skills, Inman gains the upper hand, seizing the farm tool and wielding it by instinct. Inman soon crosses paths with a wayward faithless preacher, Solomon Veasey, who has abandoned his betrothed and is about to kill a young woman whom he has gotten pregnant. Inman straps the trickster to a tree, saves the girl, and exposes the religious fraud.

Veasey, though, soon reappears, and we take closer note of him. Driven by a vision of getting rich in Texas, Veasey is recognizably American, ready to scheme and deceive to get ahead. Finding a mislaid saw beside the road, he seizes it to sell; entering a store, he brandishes a pistol to gain cash. The contrast between the two travelers appears stark when they stop at an abandoned house with beehives. Veasey wants the honey but is unwilling to labor and risk getting stung. Inman carefully takes on the task and returns with an overflowing pot and few stings. Veasey eats the honey; Inman also eats the comb, willing to take nature whole. Down the road, the travelers catch a large catfish that has swallowed a hammer. The wooden handle has dissolved in the fish's stomach, leaving the metal head. Thus we see that nature's products nourish new life but not the metal products of the age, which even a bottom-feeding catfish cannot use. Before long Inman and Veasey are captured by the novel's most evil character—the ruthless home guard leader Teague, who variously kills or sells for bounty the outliers he catches. The two men are bound with the other prisoners and led along. When he tires of tending them, Teague has the prisoners shot and pushed into a shallow grave. Veasey and the others die. Inman is gravely wounded but survives. With a primitive tool—a stone—he gains release from his metal chains.

Inman's most curious encounter comes as he enters the mountains. There he meets an older woman who lives alone, foraging, trapping birds, and raising goats. For twenty-six years the woman has inhabited a small, wheeled caravan, lodged in the same isolated spot. She is ready to move at any time but has no place to go. She fills her tiny home with books, drawings, and endless notes about goats. Despite her long immobility, she too is a rootless wanderer, committed to the road rather than to place. Her parents, she tells Inman as they sit by her stove, sold her as a youth to an older man who had buried three wives. Owner of a decent farm, he pushed her as a slave owner would dominate a field hand. Fearing she would die, like her predecessors, from "work and babies and meanness," she escaped and began living off the land. "There's a world of food growing volunteer," she informs Inman, "if you knew where to look." The woman's exposure to domination has pushed her to the extreme of isolation. It has also fueled a passionate disdain for slavery. The woman presses Inman to justify his own involvement in the war, but Inman is slow to respond. They fought to drive away the invading soldiers, he offers. After pausing he adds more. One soldier "had been north to the big cities and he said it was every feature of such places that we were fighting to prevent."

Inman is attracted to the goat woman and her freedom in the mountains. It is a life made possible, he sees, by her careful attentiveness to nature and by her willingness to let nature shape her ways. Yet Inman is troubled by her life's intense practicality, if not meanness. ("Marrying a woman for her beauty," the woman admonishes Inman, "makes no more sense than eating a bird for its singing.") Though he tries, Inman cannot imagine living alone on Cold Mountain, going months without seeing anyone.

Inman's yearning for attachment becomes stronger when he stops to get food from a young woman with an infant. The woman's small house presents a scene of elegant simplicity. The floor is scrubbed clean, the furniture is sparse, and the only ornamentation is a colorful quilt of otherworldly design. She is Sara, her husband recently dead in the war. Winter approaches, and her survival depends upon a single fat hog. Sara hungers for companionship, and her gentle touch awakens Inman to new possibilities. As morning breaks, three federal soldiers—big-city toughs, we are pointedly told—arrive and steal Sara's hog. They retreat with the hog to an old cave containing marks of prehistoric occupation (one of many signs that the struggles of the era are timeless). Inman quietly follows the soldiers and kills them. After retrieving the hog, he butchers it for Sara, allowing her life to go on. Soon Inman encounters another young woman, this one with a dead child. He constructs a casket in exchange for the food he needs to live. Then he kills and eats a bear cub when he sees that its mother has fallen off a cliff, leaving the cub to die. Thus death and life, killing and eating, dance along hand in hand.

With winter taking hold, Inman finally reaches Cold Mountain. He recognizes the terrain. Dipping his hands into a creek, he picks up a salamander, "wildly spotted in colors and patterns and unique to that one creek." The clear creek with its colorful life invites comparison with the foul tidewater rivers that Inman has crossed, inhabited by drab fish. Even sharper is the contrast between the salamander and Inman himself. Through evolution the salamander has achieved what Inman fervently seeks—a life tailored perfectly to the peculiarities of place. Inman envies the successful adaptation of this vivid animal, with its smile of "great serenity."

While Inman is working his way toward North Carolina's highest mountains, Ada and Ruby are undertaking journeys of their own, of perception and value rather than distance. They too are guided by an ideal of settled life and struggle to get to it. In their intertwined tale, Frazier extends his observations on the challenges of living right on land.

Ada has lived in the mountains for six years. She arrived in the company of her minister father, who came to regain health and lead a rural parish. When the story opens, Monroe is three months dead, but we learn much about him through flashbacks. He lived a leisurely life, sustained by bank accounts and slave-based coastal properties. He hired men to do his farm work and imported sheep to give his farm an English air. So much did Monroe admire Emerson that he named his cow Waldo ("with disregard for gender") and his horse Ralph.

We can snicker at Monroe's actions, yet behind them was a man of intellect and reason. Monroe's careful sermons drew upon diverse philosophic writings as well as scripture. His love of nature extended to all creation even as he knew that creation would not return that love. In sermons he raised the individual human to a level that surprised his rural parishioners. And even as Monroe searched for keys to a transcendent universe, he carefully studied wild things and compiled botanical notes.

Ada's adult life began in Monroe's romantic tradition. She valued nature for its beauty rather than utility. Art, music, and literature gave shape to her life. When this one-sided life comes to an end in the womblike space in the boxwoods, it is Ruby who heads her in a new direction. Hardly have they met when Ruby takes charge. Ada wonders what to do with the pesky rooster that has been bothering her. Ruby grabs the bird, snaps its neck, and proposes eating it for dinner. With the sudden snap, a new order has begun. Ruby develops work plans to bring the ill-tended farm into alignment with human needs. With Ada in tow, she inventories the farm and ranks the needed tasks in order of priority. It is a no-nonsense approach, and Ada is impressed. "Every yard of land," Ruby instructs, will be expected to "do its duty."

As Ruby works, she becomes an object of study for Ada and for us. Like the goat woman, she resists domination to the point of keeping all people at bay. She is prepared to work with Ada but not willing to live with her. Week by week she labors, schemes, barters hard, and imposes demands, all to put distance between herself and deprivation. She expresses admiration for crows, and we see in her remarks how much she resembles the clever, opportunistic birds. Also pronounced is Ruby's weak sense of progress, which we see in her responses to Ada's tales from Homer and in her comments about folk songs. Homer's tales strike Ruby as decidedly relevant. "Not much had altered in the way of things," she muses, "despite the passage of a great volume of time." As for the path of folk songs over time, the tune or lyric might change as each musician adds and subtracts before passing a song along. "But you could not say the song had improved for as was true of all human effort, there was never advancement. Everything added meant something lost, and about as often as not the thing lost was preferable to the thing gained."

Ruby's views on progress are typical of agrarian thought. So also is her disdain for distant places and other peoples' ways. Ada triggers a monologue on localism when she places an old hat of hers, imported from France, atop a crude scarecrow. Why import a hat from afar, Ruby queries, when perfectly good ones are made nearby? There was little that she needed that a person could not find or grow nearby. As for foreign travel, it too made little sense. In a proper world, people would be "so suited to their lives in their assigned places that they would have neither need nor wish to travel. . . . Folks would, out of utter contentment, choose to stay home since the failure to do so was patently the root of many ills, current and historic." Ruby's localism extends to money, which she views suspiciously. Money compares badly with the more real work of hunting and planting. Ruby is hardly unaware of monetary value, but it is use value and barter that she respects.

Ruby's efforts to bring discipline to the farm include Ada as well. She too must become productive. Ruby sticks Ada's nose in the dirt to help her learn and takes delight in displaying her spatial disorientation. The lessons that await Ada are countless. Ada, though, is an unusually apt student, and her learning is rapid. She happily absorbs Ruby's detailed knowledge even as she laments her sore muscles. Ada pays close attention to Ruby's methods of learning as well as to what Ruby knows. In a scene that displays Frazier's own localism, Ruby explains what true attentiveness entails. "You commence by trying to see what likes what, Ruby said. Which Ada interpreted to mean, Observe and understand the workings of affinity in nature." Ruby uses her ecological reasoning to speculate about dogwood, sumac, the timing of fall colors, and the ways birds spread seeds. There is method to nature, she insists. And the way to learn about it is to become humble, humble enough to study bird droppings. Ada is captivated. She openly covets Ruby's learning "in the ways living things inhabited this particular place."

Step by step Ada's learning rises, and so does her respect for nature. At one time Ada agreed with Monroe that the region's magnificent features "were simply tokens of some other world, some deeper life with a whole other existence." Now she thinks otherwise. What she sees, gazing at the land, "was no token but was all the life there is." Also changing is Ada's linear sense of time. She anticipates with pleasure the distant day when, at home on Cold Mountain, the "awful linear progress" of the years might appear to her instead as "a looping and return." In the meantime she sets higher and higher standards for her intimacy with her home ground. She aspires to identify trees by the sounds of their rustling leaves. She wants to identify wood by the smell of its smoke. Above all, she wants a vivid picture of the surrounding land and everything in it. The aim, Ada grasps, is not only to learn the land; it is also to learn how to dwell in it. "To live fully in a place all your life," Ada now sees, "you kept aiming smaller and smaller in attention to detail."

Ruby and Ada's life together is disrupted in midcourse by the surprise appearance of Ruby's father, Stobrod. We need to draw him into the inquiry also, for in his eccentricity he supplies another look at the good life. Stobrod is a natural man, stripped of culture and guided by impulse. His moves appear mostly familiar—his fondness for corn liquor, his laziness, his love of carousing, his willingness to enlist as a soldier the moment the war begins. We would write him off except for the intriguing transformation he undergoes, brought on mostly by music. Stobrod has labored at length to make a fiddle of novel design. His resulting creation includes the whittled head of a great serpent curled back against the neck. Withal, his fiddle looks "like a rare artifact from some primitive period of instrument-making." Inside the instrument he has stuck the tail of a rattlesnake caught on Cold Mountain: wildness is part of his music. For a time Stobrod played standard tunes for amusement. But during the war, the father of a fifteen-year-old dying girl asked him to play for his daughter as death came on. Stobrod quickly exhausted his small repertoire, but the girl insisted that he continue. Turning inward he began to improvise, and in doing so remade himself. What came out on that day in early 1862 was not a blend of popular elements, borrowed from songs he knew. His music came from much further back, a "frightening and awful" melody in a medieval musical mode. Stobrod had detached himself from the modern era and its Enlightenment culture. The resulting melody transfixed him, and he played it daily, tirelessly—so inexhaustible a tune, he sensed, "that he could play it every day for the rest of his life, learning something new each time."

By the time Stobrod shows up at Ada and Ruby's (his still-thieving hand trapped in the corncrib), he has linked up with a simple-minded youth who possesses uncanny ability to play the banjo. They are an unlikely pair, Stobrod and Pangle, and make uncommon music together. Stobrod wonders aloud: Perhaps the two of them might go off into the mountains to found a commune. Music will hold them together; happiness will come through freedom and self-expression. In the end their dreams are frustrated—as are Ada and Inman's—by the ruthless Teague, who shoots both and leaves Pangle dead.

Frazier's narrative comes to a rapid, intense close as Inman reaches Ada, winter sets in, and Teague and his toughs show up to wreak havoc. Random forces decide who will live and who will die. Inman's arrival scares Ruby, who attempts to keep her place by describing to Ada the kind of bountiful life the two women could have. We don't need him, Ruby argues. Meanwhile, Inman and Ada together construct their own vision of future life, and it is particularly alluring—for them and us. Both visions, of course, contrast with the ideals of Monroe, of disciplined subsistence farmers such as the Swangers, and of Stobrod and the wayward preacher, Veasey. But death intervenes, and so does new life. Teague and his troops kill Pangle and ultimately Inman. Stobrod is badly wounded. Another companion of Stobrod's, a part-Indian Georgia youth, escapes to describe the carnage to Ruby and Ada. Meanwhile Inman and Ada are together long enough for her to become pregnant. The conflicts thus resolved, we jump ahead ten years to see how life progresses on Cold Mountain. Ruby and the Georgia youth have married and borne three sons. Stobrod has stayed on with his music, and they all live on the farm with Ada and her nine-year-old daughter. The actual outcome fits none of the visions. It is merely the most realistic.

Before probing these competing visions, we need to bring together Frazier's observations about nature and humankind. They supply the building blocks for any assessment of a new, land-based order.

Nature, we see, has the power to heal when we let it—beginning in Frazier's first scene in the Confederate hospital. For humans, that healing power grows stronger when we attach ourselves to land, studying it and becoming part of it. There is vigor in sinking roots in a place, as we see again and again (most visibly in Ada's rise to health and contentment). Yet nature is often beyond our control, and we are wise to submit to it. We should woo the land rather than push it too hard. Nature is also enigmatic and mysterious, so much so that we legitimately wonder, Is there an entire other realm beyond our perception? Plainly, nature poses challenges even as it offers solace and solidity. All of these elements come together in the novel's dominant symbol—the hulk of Cold Mountain. It is distant, aloof, and solid; it offers permanence and security; it is at once a place of escape and an emblem of home; it is mysterious and cold, and yet it invites and harbors the most glorious of dreams.

As for the people, they mostly desire peace, calm, and rest. They dream of lives of no fear. They yearn for wholeness of body and spirit, for companionship, for a loving touch. Within them lies an urge to feel needed and to help others, whether they recognize it or not. Labor can bring satisfaction, particularly when it entails skill and produces sustenance and order. And there is the urge to express themselves in art or music, to exercise the imagination, to wonder about the distant and unknown. For many, the universe is frightening in its vastness and coldness. And so there comes a desire to interpret it in a way that makes humans count for something.

Along with these good human traits are the many less good ones, as Frazier vividly shows. There is the laziness, the desire to shirk labor, that we see in preacher Veasey and fiddle-playing Stobrod. It is familiar enough. In both men the trait extends to a willingness to trick and deceive to get ahead, even to steal and (in Veasey's case) reluctantly kill. Along with laziness is the inner yearning to dominate others, to bend them to one's control. That yearning, perhaps latent in many people, rises up in those who wield power. Indeed, so frequently does power corrupt that we are surprised when it does not. Military officers and the home guard are the most ruthless, but slave owners, gun toters, preachers, and many husbands and fathers are not far behind. Domination, Frazier suggests (and the goat woman declares openly), erodes the soul of those engaged in it. With few exceptions the admirable characters are those who either lack power or shy away from exercising it. Slaves, gypsies, women living alone—there are the ones most likely to act with compassion.

From strong domination it is but a short step to violence, the darkest side of human nature. Home guard Teague is Frazier's bleakest character. Joining him are nameless soldiers, including a ruthless officer who methodically kills a row of injured prisoners. Elements of human culture, it appears, can bring violent tendencies to the fore. The ruthless Teague, we can assume, feels that his killing is justified in the name of military discipline. Even clearer is the tendency of Teague's armed troops—youths, many of them—to embrace killing merely because a badge has been put on them. Early in the novel, we witness a scene of organized violence taking place on the mountains. Young men from the farms, sent to the highlands to tend livestock, encounter a band of Indian youths. For days the groups engage in a sport so violent and lawless that it is little short of a brawl. During pauses, the men drink and gamble. They have withdrawn from organized culture of any sort, and their raw natures have come through. Untended nature, we see, can be brutal. Sound culture is required to contain it.

In setting after setting, Frazier shows readers humans with the thin veneer of civilization rubbed off. Individual reason, Frazier implies, is a weak reed. A tendency toward violence often lies beneath the surface. Yet so do urges that are harmless and even creative. The transformed Stobrod, detached from the modern age, reverts to a premodern existence in his embrace of medieval musical modes. Watching him and other characters, we sense that Ruby is right: Human nature has not changed all that much over the centuries. It is culture that has changed. Unless the culture is sound, society will wind down.

Among the forces resisting a settled culture is the recurring desire of various characters to escape from daily woes to some higher realm, more peaceful and whole. Inman's yearnings remain in the world of the physical and possible, at least when he dreams of Ada and his Cold Mountain farm. But even Inman is tempted by occasional dreams of escape to another world. Monroe contemplated it in his philosophic way. Inman's youthful Indian friend (Swimmer) affirmatively sought it, as did Stobrod and others. And then there is the similar, albeit muted, desire of the novel's many orderly Christians to seek escape in heaven. How much do these Christian dreams, we wonder, differ from Veasey's escapist dream of traveling to Texas and becoming a cattle baron?

Here, then, we have the basic elements that Frazier works with as he crafts his narrative—his various ideas about nature, and the many raw elements of humankind, the good and the ill. Inevitably nature and culture will come together to shape human life on the land. And the possible combinations are many, as Frazier suggests. Which combinations seem to offer promise? Are we attracted by any of Frazier's scenes, and can we draw upon them to distill lessons for our own day?

The Life of Freedom

One choice for people, now as then, is the life of escape or (put positively) individual freedom: give in to wanderlust and the desire to live with little constraint. Frazier's narrative is populated by characters who choose this option or feel it thrust upon them. Stobrod takes to the wilds to steal corn, operate his still, and attend every party within miles—at least until his redemptive transformation. Less by choice, Ruby and the simple-minded Pangle are also alone and free when the story opens. The youths on the mountain, a band of traveling gypsies and horse traders, the killing Teague and trickster Veasey, even the goat woman in her way—all put freedom ahead of commitment. Freedom, we see, brings benefits, and is not without allure; Stobrod's music is one good fruit of detachment and independence. Yet its ill fruits are many, if only in the form of loneliness. Stobrod survived in the mountains for years, but we sense in the final scene that he is far better off embedded in a larger, settled group. The goat woman is better off alone than abused by her husband, but her prolonged isolation is costly.

Stern Agrarianism

Worth more attention is Frazier's portrayal of the Swangers—Esco and Sally—and the successful subsistence life that they sustain. There is much to admire here. No married couple appears in warmer glow than these two. They live in peace, respect their neighbors, earn their keep, and are generous in their dealings (routinely giving Ruby in trade more than she deserves). Yet, as we probe the world of the Swangers and their like-minded neighbors, we begin to have doubts. Their ordered farmstead is as dull and colorless as can be. Value is measured by human utility, and living things of no value are pushed aside. We see no hint of Monroe's love of all creation. The Swangers are not just cut off from the market; they are cut off from literature, art, and fine music. The only rural church we see is simple and severe—a clear message that God and heaven stand apart from surrounding, untidy nature. Monroe's parishioners favor repetition over novelty, sermons that merely reiterate known Bible stories and condemn sinners without prompting new thought. Their dullness and predictability will satisfy some people but hardly all. Esco Swanger's only vision of the future is peaceful but bleak: "I just want my boys home and out hoeing the bottomland while I sit on the porch and holler Good job everytime that clock strikes the half hour."

The agrarian world of the Swangers shows an even darker side when Ada recalls a party at their house on the eve of the war. Farm women of various ages have gathered to chat. An older women offers news of her daughter, who has married poorly. Dominated by her husband, the young woman shares a house with a family of hounds who lounge in the kitchen and leave dog hair in the food. The woman has borne children year after year and now views matrimony "as a state summing up little more than wiping tails." The women laugh. Ada, in contrast, "felt for a moment as if she could not catch her breath."

What we discern, from these incidents and others, is that the agrarian culture of the Swangers largely accepts the domination of women. It is also prone to condemn people for their mistakes and leave them to their ill fates. A woman who chooses a mate poorly is laughed at, not assisted. Ada has more humane views, made possible presumably by the free thinking of her father and her access to larger worlds. She can critique the patriarchal oppression of the age. She can see, too, the possibility of transformation. When Stobrod returns to Ruby's life, Ruby is prone to condemn him for past transgressions. Ada in contrast is more forgiving, more impressed by the change that has taken place within him. Ada's generous spirit is a useful advance over the disciplined, judgmental world that the Swangers inhabit.

Two final pieces of Frazier's narrative help fill out our understanding of the land-based culture that the Swangers represent. As Ada approaches the Swangers' farm on her day of reawakening, she passes along the road a man with a stick, beating a bag of beans hung from a tree. As he labors he curses at the bag as if it is an obstacle to his desired life of ease. The man is beating the shells off the beans, he tells Ada, and every bean, he relates, "was a thing to hate." All of his work on the beans—plowing, planting, tending, weeding—had been done in hate. Moments later Ada reaches the Swangers on their porch. They too are working beans—threading them on strings to hang up to dry. Their attitude could not be more different. Esco and Sally carefully work the beans, slowly and quietly, touching "each pod as it if were a thing requiring great tenderness." (Beans appear repeatedly in Frazier's tale—reminding us, again, of Thoreau's experiences—and we can largely judge characters by the ways they relate to them.) We are thus left to wonder, If the kindly worldview of the gentle Swangers can be so severe and judgmental, what kind of world would we have if the bean-beating farmer were our agrarian norm?

Finally, our assessment of the Swangers must draw in the simple-minded Pangle, left at a young age to roam the mountains alone. He is called Pangle by Stobrod because he seems to resemble a family of that name, but Stobrod wonders whether he might be a Swanger instead. How could the Swangers, though, have had such a child and cast him off? And might the Pangle family be just as admirable as the Swangers? The question is essential, for it leaves us wondering, How well does the agrarian world of the Swangers take care of misfits? In a world of independent farmers, what happens to people who cannot stand up alone? And what about the people who rebel against the discipline or simply get bored?

Liberal Transcendentalism

From the Swangers' subsistence farm, we can return briefly to Monroe and his transcendental ideals. His life, plainly, is not sustainable on its own terms, given its dependence on bank accounts and distant plantation labor. He lives on a farm but is hardly rooted in it. He is as detached from nature and committed to the money economy as many of the landless characters. By the time the novel opens, three months after his death, his money is exhausted: he would have been poor. Still, we can admire Monroe's moral vision, which broadly embraces nature and which attaches significance to each individual human. Monroe's questioning, his attentiveness to reason, his engagement with ideas, his worldliness, his embrace of art and music—these and other traits have appeal. His emphasis on natural aesthetics is perhaps excessive, but it highlights all the more the deficiencies of his neighbors. Monroe is a man of imagination, and we respect him for it. What is troubling is his tendency to detach too much from the physical to contemplate the metaphysical. Yes, we should recognize the limits of human knowledge about the world, but Monroe's speculations appear too fanciful. Related to this limitation is another one, highlighted by a comment Ada offers late in the narrative. By then, her grounding in nature has enabled her to think more critically about her cultured childhood. Inman has arrived, and Ada is anxious to describe her cultural journey since Monroe's death. Ada mentions the knowledge she has gained about the land and its many living creatures and forces. Along with this good news, she has "one terrible thing" to relate: her recognition that Monroe "had tried to keep her a child." Monroe had failings as a father. Though a free thinker, he had not set his daughter free. He pushed her into the Victorian parlor, cut off from the realities of life.

The Pursuit of Plenty

Monroe's vision is so obviously incomplete that it cannot stand as a developed option. More serious is the vision that Ruby presents when she fears that Ada will turn away from her and take up with Inman. Ruby largely embraces the worldview of the Swangers, so we need not dwell upon her case long. What requires our attention are the alluring details that Ruby includes. When it is clear that Ada wants Inman in the picture and that Ruby herself will not get evicted, Ruby presents her vision to Ada, scratching in the dirt with a stick as she talks: "She put in the road and the house and the barn, scratched up areas to show current fields, woodlots, the orchard. Then she talked, and her vision was one of plenty and how to get there. Trade for a team of mules. Reclaim the old fields from ragweed and sumac. Establish new vegetable gardens. Break a little more new ground. . . . Years and years of work. But they would one day see the fields standing high in summer with crops."

Ruby's vision of plenty is vivid and attractive, but it fails to address key points. We wonder about the people who will live here and whether they, like Ruby, can be satisfied simply by having plenty to eat. We wonder about the relations among them—husband and wife, parent and child—and about whether her agrarian order has appropriate ways to contain and satisfy the full range of human traits. Perhaps most plainly, we wonder what limits Ruby herself will respect as she goes about reclaiming old fields (the ones that Monroe let go), breaking new ground, and enlarging the orchard. Are there limits to this cutting, plowing, and replacing? Will she continue pushing nature harder and harder to produce ever more? Earlier, Ruby explained that she wanted to get rid of Stobrod's fiddle—as a way of getting him to work, we can assume. She was particularly offended by the rattlesnake tail (the wildness) in it. For her, nature needs disciplining. We must wonder whether she will take her discipline too hard. What will happen if one or more of her family members turn against her? And when "every yard of land" must do its productive duty, what will happen to the ecological functioning of the landscape?

A Marriage of Nature and Culture

No sooner has Ruby finished drawing in the dirt than we overhear Ada and Inman as they piece together a different vision of the farm. It too deserves attention. The two lovers have broader dreams. Inman recalls seeing a portable sawmill and thinks he might buy one, hauling it from land to land and sawing out the material for each landowner to build a house. "There would be an economy in that, and a satisfaction for the man as well, for he could sit in his completed house and delight in all its parts coming right from his own land." Inman is prepared to join the money economy, yet he recognizes the stability and confidence that come from direct production from land. A sound economy is vital, and so is the satisfaction of laboring to produce from nature. This economy, Inman says, would make much possible: "They would order books on many topics: agriculture, art, botany, travel. They would take up musical instruments, fiddle and guitar or perhaps the mandolin. Should Stobrod live, he could teach them. And Inman aspired to learn Greek. That would be quite a thing to know. With it, he could continue the efforts of Balis." What they propose, then, is a full marriage of nature and culture, of a type neither achieved nor even contemplated by any other character. They would draw upon the fine products of the world—a simple shotgun, fishing tackle, and boxes of watercolors from England, for instance—but with the aim of developing a settled, contented existence in the shade of Cold Mountain. And they would grow old together "measuring time by the life spans of a succession of speckled bird dogs." Ada, we see, has not rejected her father's Anglican culture. She has drawn upon it sensibly and then attached it to the local land, using it to enrich an otherwise dull and confining existence. Inman's strong link to nature has not left him uninterested in the scholarly culture of his former hospital companion.

Ada and Inman get no further than their dreams, for Inman soon dies in a bloody shootout with Teague and his colleagues. Inman is close to killing or scattering all of his attackers, Teague included, when he pauses before killing a young man who has tried to kill him. Inman tries to talk sense to the youth, encouraging him to put down his gun and walk away. With the flash of a hand, the youth ends Inman's life.

Ten years go by, and we receive an update. Ada and her daughter are there, and so are Ruby, her husband (who worked as a hand for two years), and their three sons. And Stobrod is there, too, disciplined enough to help with farm chores but contributing even more with his fiddle and his compositions. Ruby appears to manage the work, but it is Ada who stands high, as owner of the farm and guiding light. Many clues let us know that she has succeeded in immersing herself in local nature. Monroe's broad concept of value continues to shape her life. Cold Mountain has become for her a particular object of study. She watches closely as autumn proceeds and the fall colors day by day move down the mountain and overtake the green. In the evening Stobrod plays and Ada's daughter sings with a clear, strong voice. Books are also part of their life.

Ada reads a tale from Ovid about two old lovers who live together so long in peace and harmony that they turn into spreading trees. Ada's lover, of course, is absent. Yet their dream of permanence is deeply rooted.

Frazier's novel is particularly useful because it brings life into sharp focus, exposing its key elements and inviting readers to engage with it. We can sense his judgments but are able to reach our own. *Cold Mountain* is a story of love among the ruins, a tale of new life begun by a surviving remnant after destruction has wound down. The basic narrative line is easy to distill, in terms of the possibilities for new life on Cold Mountain. The Swangers and their subsistence farming provide the thesis; Monroe and his cultured transcendentalism provide the antithesis; Ada and Inman's vision offers a synthesis. With Inman's death we do not get to that synthesis, but the ultimate resolution comes close. What are the main elements of this synthesis? Do they include raw materials for criticizing and reforming modern culture? Do they provide building blocks for an enduring life on land?

An enduring society, Frazier suggests, must attend first to nature. It must study land carefully and develop a fondness for it, all of it. It must adapt itself to nature, not just in terms of its economy but in its aesthetic standards and its affections. Ada provides the ideal, on this point and others. We live best in a place when we are endlessly fascinated by it. Monroe was right in his broad sense of value in nature. Ada wisely maintains it, even as she absorbs Ruby's practical knowledge.

Just as important are our interactions with one another. There must be honesty in them and freedom from domination. Settled life is not a solitary activity; it is a shared enterprise. Good land use requires a successful social order, including mutual respect within the family. That order, in turn, is more likely to arise among people who live simply and who take pleasure in the common elements around them. On this point, the Swangers provide good guidance. Simple wants are easier to satisfy.

Our use of nature, we see, flourishes when we work with nature rather than against it, wooing the land and using its natural processes. Inman witnesses an example of this ideal early in the novel. He has paused during his initial visit to town to buy clothes. Looking skyward, he spots a circle of swirling vultures. As he watches, the birds do not strike a wing beat "but nonetheless climbed gradually, riding a rising column of air, circling higher and higher until they were little dashes of black on the sky." The birds take advantage of nature's powerful forces. Humans could do the same. The wise vultures provide a counterpoint for incidents that unfold as Inman journeys—scenes of angry people who fight against nature and either lose or end up laboring too hard. Frazier leaves no doubt: Nature shows no concern for humans, and humans can make it better. But they can just as easily degrade it.

Stobrod's lingering presence reminds us that the settled, land-based life works best when people can express themselves in art, music, and story. They must have ways to spin hopes, offer dreams, and stretch imaginations. Without them, how can the human soul flourish and remain whole? And then there is the matter of wildness, and the need to keep it close at hand—even within a musical instrument.

The most difficult piece is the need to control human deviance, to keep misfits from degrading society and running off to cause trouble. On this point Ruby has aided Ada considerably and continues to assert control. Stobrod, we see, has become productive without losing his music. The Georgia youth has also given up his wandering and taken to the settled life, disciplined by Ruby's tough love. In our brief scene of Ada and her daughter, we see Ada quick to intervene when her daughter pulls a lit stick from the fire and waves it dangerously. The girl complains; Ada remains firm. For the moment, the social order is intact; culture controls rowdy nature. But three sons are growing up, and we cannot know what lies ahead as they itch to engage the world.

Ultimately, Frazier leaves us at a sound if necessarily incomplete place. Our problems living on land are chiefly problems of human behavior and culture. They are problems that relate to our values and to the ways we interact with one another. This is where we must focus our efforts to live better. Can we rise about our sinfulness, in the here and now, on earth? When hate is so easy, can we love the world as it is? Can we anticipate redemption and work to bring it about? These are our questions.

Frazier's closing scene, ending his narrative, proposes a new beginning for the human enterprise. With Inman dead, Ada stands as the female Adam. She will lead the way, our fullest embodiment of nature and culture and thus our best hope (worthier by far than Scarlett O'Hara). If the enterprise succeeds, it will be because of her and the synthesis she has achieved. Ruby is there to help. Our society, then, is matriarchal, and hardly by accident. The culture of domination must give way, and a new order must arise. Yet we must remember the hard route that Ada took to get to this lofty place. Her old life had to end; she required a rebirth out of nature. She had to learn a new language—the language of the land—and to be humble as she did so. She needed to set aside the broader world to dwell on the practical possibility of enduring life in a particular, intimately known place. That was her education. Could it be ours?

4

FRAMING OUR CHOICES

To judge from its popularity, the journalistic convention of reporting two sides to every controversy reflects something like a deep-rooted yearning—certainly among Americans, perhaps among other peoples too—to reduce complex issues into opposing options. The world is hardly so simple, of course. Dichotomies are as apt to confuse as they are to clarify. Still, we routinely find ourselves confronted, for instance, with the options of vast reproductive autonomy for women or a fetus's unlimited right to life: pro-choice or pro-life. Both sides ignore the complex social world in which abortion decisions are made. This tendency toward two-sided simplification is hardly new. When Abraham Lincoln confronted Stephen Douglas over slavery, he chose to frame the enigma simply in terms of moral right and wrong, as if slavery did not also pose knotty practical problems of land use, economics, suffrage, power, and daily living. These practical issues became painfully evident after the war, as legal freedom for blacks meant less and less in practical terms. Despite its flaws, the rhetoric of two-sides-to-every-issue freely roams the social landscape, casting its usual mixture of seeds.

When it comes to talking about people, nature, and land-use conflicts, three dichotomies seem to dominate today's rhetorical scene, sometimes overtly, sometimes by implication. As with most dichotomies, the choices they pose are simplistic and incomplete. More troubling, they are also misleading if not false, so much so that the dichotomies themselves—the ways we frame our land-related choices—have become a major part of our ecological quandary.

Preeminent among the three popular dichotomies is probably the claimed split between protecting nature and taking care of people. A familiar expression of the division goes like this: Environmentalists care only about nature, not about humans (or jobs, or communities, or the like). Shall we satisfy human needs for oil or instead protect the Arctic caribou? Shall we keep our polluting factories rolling along, paychecks flowing, or should we shut them down in the name of protecting fish and birds? Nature versus people: those are the basic options, so we are told.

So inaccurate is this dichotomy that we ought to view its endurance as a bad cultural omen. For decades environmental activism has aimed above all to protect human health—clean air, clean water, healthy food—and to provide natural places for people to go and wild things for them to see. If the environmental movement is forced to reside on one side of this miscast dichotomy, surely it lives on the human side, where pretty much everyone else stands. A greater flaw here is that humans are embedded in nature and dependent ultimately on its healthy functioning. Humans cannot survive apart from nature. Only by protecting nature can we really support people. Thus the entire dichotomy is false in its assumption that we can somehow promote humans while ignoring nature. Exposed to the light of ecology, it withers.

Equally miscast is the familiar assertion that our current policy choices align along some liberal-versus-conservative spectrum. The pro-environment side of these questions is dubbed liberal. The policy stance that cares less for nature, perhaps by process of elimination, is viewed as conservative. As for the latter, it is not particularly clear what the conservative side is out to conserve. Certainly it is not other life forms, natural habitats, or the options that future generations will enjoy. The likely targets of conservative protection would seem to be our inherited ways of perceiving land as a limitless resource pile, awaiting our consumption, along with our patterns of disregarding basic processes. Thus we act conservatively, it appears, when we unleash the bulldozers; we act conservatively when we fail to conserve. On its side, liberalism originally entailed a spirit of generosity and openness. That definition long ago gave way in political terminology to a perspective that elevates the independent individual and downplays organic social structures. In the liberal view, individuals ought to enjoy freedom to make their own life choices, liberated from unfair shackles. When "liberal" is used in this classic sense, of exalting individual initiative (in terms of lifestyles and economic activities), both American political parties are basically liberal. They differ chiefly in that the Republican Party stresses economic liberalism while the Democratic Party is inclined to promote social liberalism. In both parties, it is the individual who is supreme.

This definition of liberalism—derived from eighteenth- and nineteenth-century British political writing—has taken a beating of late. The term's meaning in common speech has drifted, leading to semantic incoherence. For many, "liberalism" has become a popular synonym for moral laxity, bad judgment, and impracticality. As for "conservative," that term first gained positive connotations only in the 1950s; until then it was a term of derision, mostly connoting backwardness. That was about the time when political conservatives turned away from the classic ideals of Burke, Carlyle, and Richard Weaver and became friends with the spirited defenders of the unfettered market. (Earlier conservatives like Burke had frequently attacked the market.) Since the 1950s, conservatives have gained vast political power by becoming more economically liberal.

To the extent the terms "liberal" and "conservative" possess much meaning today, the dichotomy that they pose is useless (or even worse) as a way to frame our choices concerning land use and nature. If liberalism means maximum individual liberty to act as one pleases, with little regard for ecological ripple effects, then environmentalism stands on the other side. When conservatism means maximum freedom for corporations to carve up nature and put it on the auction block, without regard for nonmarket values and the future, then environmentalism is opposed to it as well. On the other hand, when and if liberalism promotes overall health, ecologically defined, then environmentalism will listen. And to the extent conservatives demand communally responsible behavior (by polluters and landowners as well as street criminals), then environmentalism will also perk up. As things stand now, though, neither liberals nor conservatives seem to care much about land as an integrated whole. Neither side pays attention to people who are laboring to use land well.

The linguistic drift of the terms "liberal" and "conservative" seems to be repositioning them so that they now overlap with a third dichotomy, which is also used to frame environmental stories. This is the choice between individual freedom and governmental coercion. Liberalism, one frequently hears, is about government telling people how to live their lives, rather than allowing them to make their own choices. Conservatives, it is said, are more likely to let people decide for themselves. Because environmentalists are prone to place limits on what people do, it would seem to deserve the liberal label. It is situated on the coercion side of things, with conservatism presumably on the other side.

Here, too, we run into massive confusion and factual inaccuracy. Government is not the only source of coercion. People are also coerced by private actors, particularly those with wealth. And quite often, the private actors are wielding power that is vested in them by laws and government. If liberty in fact is our prime political value (as conservatives say in economic matters and liberals say in social matters), then we need to ask, What about the liberty to breathe clean air or to drink clean water? What about the liberties to enjoy wildlife and to rest content knowing that one's daily living practices do not sap the long-term health of the land? When do these liberties gain protection?

Liberty comes in many forms. It has a positive side as well as a negative one: it is freedom *to* as well as freedom *from*. And it is collective as well as individual: freedom to act *together* as well as *alone*. No political liberty is more vital than the liberty of one citizen to join with other citizens to make rules and plans to promote the common good—the liberty celebrated in the Declaration of Independence. Indeed, many public-policy options (restoring a major river, for instance) become feasible only when citizens pursue them in concert rather than acting alone. When we curtail government's powers in the name of protecting the negative liberty of individuals, we necessarily foreclose particular land-use options that require landscape-scale planning, in the process reducing positive and collective forms of liberty. When it comes to action by government, in short, liberty resides on both sides.

As for the coercion side of this proposed dichotomy, we forget that private property is itself built upon coercion. The landowner who posts No Trespassing signs draws upon public power and police to restrict the liberties of the wandering public. To have the legal power to pollute and degrade nature is, in effect, to be empowered by law to harm other people, without their consent. Also, to degrade land is to impose limits on future generations, as well as other life forms. Liberty and coercion are bonded twins, not real opposites. To create a sphere of liberty for one person—whether it is liberty to sleep unmolested at night or to clear-cut a forest—we must necessarily curtail the liberties of other people. Rachel Carson raised this issue about liberty in her attack on pesticide-use practices. Americans were having their lands and even their bodies subjected to pesticide spraying, without their consent and with little opportunity to learn about pesticides and set rules for their use. Their liberties were under attack, Carson claimed. In this view of affairs, the environmental side of the issue is the home of individual liberty. Polluters and pesticide users are the ones who stand on the side of coercion.

Given the grave flaws in these three prevailing dichotomies, there is an urge to say that we should simply get rid of them. Then we could deal with complex issues in their full complexity. Inevitable tensions and paradoxes underlie all land-use problems. Rarely are they captured by the kind of moral distillation that Abraham Lincoln found politically convenient in his battle with slavery.

Alternatively—and here we get to an option that is more likely given our temperaments—we could attempt to frame our land-use choices not by using a single dichotomy but instead by using several of them at once, and better ones than we now have. It is possible, in fact, to craft useful, framing dichotomies having to do with people and nature. No one of them alone would capture more than a piece of things. But several together could supply us much of the clarity and complexity we now lack. These dichotomies could operate like multiple cameras, allowing us to consider our land-use predicament from differing angles.

Here are seven possibilities. Used together, they could help clarify where things stand in terms of nature and American culture while shedding light on paths ahead.

Autonomous Self versus Embedded Self

Many disputes about using nature build upon an underlying disagreement about the best way to understand people. Are they best understood as autonomous individuals, able to chart their own paths and succeed on their own initiative? Or are they more accurately viewed as embedded in social and natural orders, with their successes and welfares intertwined with the flourishing of these orders? Other apt phrasings are also possible: Are well-functioning neighborhoods and communities important or unimportant to individual health and success? Is governance best undertaken by toting up the preferences of autonomous individuals, or do better decisions emerge when people form a collective group and work together as group members? These are all fundamental questions having to do with ontology. They challenge directly the self-guided individual as overall cultural ideal. Are we all merely individuals, interacting by choice, or are we bound together in ways that entail dependency? If the latter, do we diminish ourselves and our future options when we undervalue our interconnections?

This dichotomy has real and useful substance. On its side, environmentalism is about identifying and honoring connections and dependencies. The opposing view, when it comes to nature and future options, is inclined to ignore them. We have a clash.

Nature as Parts versus Nature as a Whole

Drawing upon ecological principles, the environmental perspective sees nature as an integrated whole. Nature is held together by ecological processes that are essential to its overall functioning and thus to the health of its living members. People certainly use this natural whole, but we are wise, when doing so, to respect its functional integrity. In the opposing view, nature is basically a warehouse of raw materials—a vast gathering of colocated parts—some of them valuable, most of them not. In this more atomistic view, nature's parts are valued chiefly in isolation, usually through market processes that aggregate individual preferences. Nature's parts are also fungible in the sense that, when one of them (one resource) runs out, we can find another to meet the same human need. Market incentives will lead clever people to find substitutes.

The conservation movement that arose early in the twentieth century largely embraced this view of nature as parts (resources). It insisted only and importantly that flows of the valuable parts be maintained over the long term. Only later, beginning in the 1920s and 1930s, did leading students of humans-and-nature spot the flaws in this resource-focused reasoning. They began shifting to an ecologically integrated view of good land use. Today, a rather clear dichotomy exists, and we should mark it clearly. Is nature valuable as an integrated functioning whole that we should respect, or is it instead a collection of parts that we can turn over to the market to manage? Another clear clash.

Present Generation versus Future Generations

Many land-related disputes raise questions as to the best temporal perspective to use. Should we think only about people living today, discounting all future harms and benefits? Instead, should we afford future generations independent respect? Put otherwise and perhaps in more familiar terms: Are we morally obliged to protect nature and preserve land-use options for people not yet born? Are we subject to religious duties that command respect for creation?

Popular environmental rhetoric often includes the adage (attributed to Iroquois tribes) that we ought to plan ahead for the seventh generation. Something like the same idea is incorporated into the vague ideal of sustainability. On the other side—among people who reject any such duty to future generations—the claim is made that the best way to plan for the future is to allow the market to operate with little restraint. Market processes, it is said, will create economic incentives to protect nature to the appropriate degree. This reasoning, we should note, is easily knocked down, even if we accept the various normative values that underlie it. For many reasons well described in economic literature, the market is not up to the task of using land in ecologically sound ways.

We do not need to engage the merits of this issue to see that we have a third dichotomy, which also helps lay out our options: do we allow individuals today to use up nature, deranging it or protecting it as they see fit, or instead should we collectively and deliberately take steps to preserve nature for the long term?

Partial Accounting versus Full Accounting

Environmental thought is guided by principles of interconnection—in nature, among people, and among generations. From this perspective, environmental thought criticizes the ways we typically describe and evaluate our land-use options. When we calculate costs and benefits, environmentalists claim, we fail to trace the ripple effects of a land use far enough to see where they lead. We fail to identify all the relevant costs and benefits. This complaint highlights another useful dichotomy. One side presses for a more full accounting. The other side is largely content with a partial accounting that pays attention only to what is obvious.

This conflict crops up, for instance, when we hear about how costly it would be to take good care of land, in terms of preserving species and ecological processes. The problem here, environmentalists say, is that cost estimates for protecting nature are frequently calculated in ways that ignore the often greater costs of *not* taking care of nature. Or they ignore the fact that losses to one person or in one place are often directly offset (or exceeded) by gains that accrue to another person or in another place. For instance, when forests in Oregon are protected and jobs decline, new jobs could well arise in the forests of Georgia. How sensible then is it to calculate the costs of protecting the Oregon forest by considering the resulting job losses locally but not the resulting job gains farther away? Not very, but we do it. Similarly, we readily ignore how intact forests protect waterways, thereby sustaining fisheries, lowering the costs of purifying drinking water, and adding value to waterfront properties. Economists often talk about the externalities of land use, but they often understate their frequency and complexity. In their cost-benefit analyses, they rarely trace externalities very far and often ignore many subtle ones.[1] Thus, when a former prairie is turned into a cornfield, the annual corn harvest is viewed as pure income, as if we gained nothing economically from the prairie and suffer no loss now that it is gone.

Here too we have a distinct split of policy views, at least at the poles. One side aspires to assess land-use decisions at broad spatial scales, considering the long term and including the full range of tangible and intangible land uses. The other side typically thinks about land-use issues on a parcel-by-parcel basis. It mostly asks how much money various options will make for the owner. In one view, landowners have valued neighbors in space and time, near and far. In the other view, neighbors play little or no role in the calculations unless they are harmed in glaring ways.

Faith in Information versus Faith in Sound Values

This next dichotomy has several elements to it, having to do with our knowledge about nature and our abilities to reason. We can state the elements as follows:

In one view, people can act sensibly toward nature based on what they know while pretty much ignoring everything else. If our knowledge is not yet complete, at least it is close enough. In the opposing view, we should act more humbly, recognizing our penchant to err and the inevitable limits on what we know. Caution is wise, and collectively we should embrace it.

In one view, we can wipe landscapes clean and reconstruct them from scratch, bringing in the few organisms (often genetically altered) that we prefer and keeping out all others. (This is the atomistic paradigm of modern farming and, increasingly, modern forestry, aquaculture, and landscape design.) In the opposing view, we are wise to keep as many of nature's parts as possible, on the assumption that the organisms and natural communities that have survived in a place for millions of years might just have wisdom within them that we ought to respect.

In one view, our environmental problems are caused chiefly by a lack of information, which means the solution is to gain more facts. In the opposing view, facts are inevitably incomplete, and we should place our hope in sound cultural values. Ignorance is our fate, and our decision-making processes should reflect it.

In one view, we can rightly charge ahead, altering nature in ways that we cannot readily undo, leaving it to clever people down the road to correct our missteps. In the opposing view, we should avoid altering nature in such ways in the event we realize that we have erred.

In one view, people have arisen to displace God as manager and creator of all that we survey. In the opposing view, hubris of this sort is foolish and immoral, if not criminal.

Market versus Democracy

Mechanisms of some sort are inevitably required to make collective decisions about nature and about how we use it. The more complex and integrated a society, the harder these collective decisions are to make well. One influential decision-making mechanism is the market. It collects preferences of individual consumers (one dollar, one vote) and translates them into resource-allocation decisions, shifting resources to their "highest and best uses." Another decision-making method is for people to assemble as citizens and make decisions (individually or through their representatives) about their shared landscapes. Democracy has seen better days and may see them again. Certainly we need it to. These, though, are the two basic approaches we can use. To identify them is to produce another useful dichotomy. When it comes to using land, do we need more market or more democracy?

The market's failings are many, according to those who study markets. Also, many land-use goals are simply not achievable without coordination of a type that the market frustrates. Many of the market's current failures could be corrected by imposing legal restraints on it, but such legal limits can arise only out of democratic processes. This means they are feasible only if we first find ways to make democracy work better. In short, even if we prefer the market-property-rights approach, it takes sound democracy to create new property rights and define them legally in the right way.

We should not overstate the difference between these two decision-making mechanisms. In fact, they overlap quite a bit. The market works only because laws and democratic public institutions support it. Democratic governance, in turn, often relies on market mechanisms to achieve or implement publicly set goals. Still, the dichotomy is usefully added to the list. On one side are those who have faith in the market, at times viewing it as a savior for all land-use ills. On the other side are those who see the unbridled market and the ideology that it employs (chiefly nature-as-commodity reasoning) as a powerful engine of environmental decline. One side says we need more market; the other side, more democracy, of a strong (that is to say, citizen-run and deliberative) type.

Faith in Tools versus Faith in Culture

To have faith in the market is to believe that a well-designed tool (an economic mechanism) can somehow transform vice into virtue. Market mechanisms, we hear, can lead invisibly to sound land-use outcomes, even when people hardly care about nature. This policy approach exemplifies a more general tendency, to believe that we can solve our environmental and land-use ills simply by coming up with some new tool to remedy them. With luck, the tool will require no real shift in us and what we know and value.

This tendency to search for tools and talk about their relative virtues shows up clearly in much academic and quasi-academic writing about the environment. Academics line up behind their favored tools, often touted in technical ways. Thus we have proponents of cost-benefit analyses (including shadow pricing) who imply that their tool can bring us good landscapes. Others want to carve all of nature into pieces and turn each piece over to a private owner; privatization, we hear, is the best of all tools. Improved agency-administrative processes also have their adherents (with new and improved processes, this reasoning goes, governments will make better decisions). And then there is the talk about the many likely benefits of putting a price tag on nature's "ecological services." Perhaps we should issue trading rights in environmental "bads" or "goods" and foster markets in them. Or perhaps instead we should promote "ecosystem management" and bottom-up, community-based conservation measures. Or perhaps the real key is to tout the virtues of "cradle-to-grave" industrial processes.

Embedded in each of the tools is an element of practical wisdom. But none of the tools standing alone are likely to accomplish very much. They simply do not get to the root cultural causes of our land-use ills. No tools, however clever, can substitute for a culture that respects nature. No tool is adequate unless people earnestly seek to promote the land's lasting health.

Aldo Leopold expressed this ultimate view when he asserted that conservation was not something that a nation bought, it was something that a nation had to learn. Leopold aimed his barb at conservationists who proposed improving landscapes by having government buy more land and by paying private landowners to act better. There was no need, the money-dispensing conservationists seemed to say, to tinker with the ethics and laws of land use. We can apply Leopold's criticism even more broadly than he did. New tools can help promote conservation, to be sure. But people select the tools and wield them, when and as they see fit. And it takes sound values and a clear vision of good land use to craft the tools to begin with. So how can we get good tools and use them well unless we first have a well-rooted culture? The choice of tools, logically, would seem to come second, after we have good culture in place. If bad culture provides the foundation for bad land use, then it is good culture that offers the hope of real change.

To identify these seven land-related dichotomies is hardly to chart a clear path ahead. Indeed, these dichotomies raise knotty questions without really answering them. They pose options and clarify what is at stake. Still, we can hope that greater clarity of thought can move us in the right direction.

What if, starting today, journalists employed these dichotomies in their writings, instead of the tired, misleading dichotomies they use over and over? Would new language in the public arena encourage people to question their values and consider new ones? At the moment, we do not know.

Note

1. The deficiencies in economic analyses are far greater than this sentence suggests, even in terms of analyses of externalities (that is, how we think about the links between one land parcel and surrounding ones). In the view of economics (which as commonly understood slants strongly against the good care of nature), externalities skew the economic calculations of landowners, and thus the fewer of them the better. From an environmental perspective, lands and thus landowners are integrated in natural systems. Good land use fosters healthy links among landowners, just as it does between present and future owners, while paying attention to the vast limits on our knowledge of these links and on our ability to trace and evaluate them.

5

Good-bye to the Public-Private Divide

To live well on land has long been a challenge and a hope for people everywhere. It is the "oldest task in human history," Aldo Leopold claimed, and he was in a position to know as a careful student of the land and of the ways various peoples had misused it.[1] In America today, we are having trouble at that task, according to many conservationists. A major cause of our troubles has been the institution of private property rights in land. Too many landowners use their lands in ways that undercut the collective good, and their property rights shield them from accountability. Particularly in the American West, we hear another complaint about landownership, having to do with the massive federal land holdings. Federally owned lands are also being misused, many allege. Some say too many public lands are off-limits to the kinds of extractive land uses that produce jobs. Others contend that publicly owned lands should serve public purposes alone, and that the public's prime needs are to promote wild species and ecological processes while supplying places for recreation.

I want to address this subject of landownership, with particular regard for the division between private and public lands. Given how lands are mingled, private with public, it is difficult to talk about one form of ownership except in relation to the other. So after exploring the institution of ownership generally, I propose to set these forms of ownership side by side to see how different they really are, asking why the two forms exist and whether the future of one form of ownership might depend closely upon the future of the other. Is it possible that the problems with one ownership form are linked to the problems with the other? Indeed, is it possible that the simple division of lands between private and public is itself a problem? The answers to these questions have relevance to all lands everywhere.

The place to begin is with the private form of ownership. We need to pry open the institution of private rights in land and look at its inner workings. If we can do that, probing why private property exists and what it is supposed to accomplish, we can gain a sense of how property has changed over time and where it is heading today. Armed with that understanding, we can then turn to public ownership, to figure out how public land differs and why it too exists.

To start, let us set aside essentially everything that we know about landownership and begin simply with the land itself, a natural scene. Imagine a valley somewhere, vast in extent and empty of people. Insert a river, meandering through the scene, along with a few hills or mountains, some patches of trees, some wildlife. It is a good place to live, with reasonably fertile soil, maybe a fair amount of rain, some timber and rock for building. Nature is at work, with its cycles of wind and water, of birth and death, of nutrients coursing through the system, and of plants and animals that, in their ceaseless competition, have formed a resilient biotic community.

Now let us add people to the picture, perhaps perched on a hillside, looking out over the plain. These people have arrived from afar and plan to stay, settling in and making their homes. To do that, they obviously have to use the land. Perhaps they will not have many troubles as they go about their work, if the land is abundant and reasonably uniform in its attractiveness. But these assumptions are not realistic, so let us modify them. Let us assume the land is expansive but differs widely in its natural features. Some places are far better than others to build homes. Some places are rich in wildlife, or have more fertile soil or bountiful grasses. Some lands are next to the river and have good water, while others are higher and drier.

These arriving people face a question. How are they to organize themselves so as to use the landscape successfully? If person A takes over one tract of land, making exclusive use of it, then other people will be unable to use the tract. That is, if we let A claim ownership over a particular piece of land, we have necessarily limited the ability, or we might say the liberty, of everyone else to use it. When everyone can use all land freely, the liberty of all is equal. But the moment we give A special control over a tract of land, then we have done two things: we have increased the liberty of A, and we have decreased the liberty of everyone else.

Back to the question: how might the people organize their affairs to make effective use of the lands? The question is difficult and the possible answers countless. The people could divide the land into numerous small pieces, or they might instead keep the land undivided. They might use the land by laboring in teams, or they might use it as individuals. A particular tract of land could end up not with one user but with several people holding use rights in it. One person might gain the right to graze animals, for instance, while someone else held rights to use the timber, hunt wild animals, extract water, or merely walk across the land. Use rights could go to families instead of individuals. They could be limited in duration or unlimited. Perhaps some places will be set aside and not actively used by anyone. To add to the complexity, let us recognize that one person's land uses can easily disrupt the activities of other people, and so there are countless questions about how the use rights of A fit together with the use rights of B and how the ensuing conflicts will be resolved.

As the people go about deciding how to use this bountiful land, they will no doubt consider the human side of the issue—their needs for food, fiber, and shelter, as well as their desires for recreation and social interaction. Some needs are basic to all people, but many needs will depend upon the peculiarities of the arriving people, including their social values and structures, their religious beliefs, their senses of individual autonomy and equality, how much they value privacy, what weight they give to future generations, and so on. Along with these human needs will be the many factors that relate to nature itself, to the variations in the land and its ecological functioning. Some lands will tolerate human use without much effect; other lands will not. Some lands will have special value in supporting wildlife or sustaining ecological processes. Good land use will take these natural variations into account.

As the people think about their work, they will be wise to explore all of these factors. Even so, they will make mistakes. Much about the land's natural features and functions will be unknown or misunderstood. As for the people, their numbers no doubt will change, and so will their technology, their values, and their dreams. Patterns of land use that make good sense at one time might not make good sense years later. Change is inevitable, on both the human and the natural sides. As people alter the land, they may come to see it differently. Parts of nature they once viewed as common or unimportant may become scarce or otherwise highly valued. If the people are particularly wise, they will anticipate such changes by crafting mechanisms to adjust their patterns of owning and using land over time.

Let us set this scene aside and turn to three others, which we can sketch more quickly.

Scene 1: Hunter Albert for years has used a vast forest to find game for his family's table. He is a skilled hunter and knows animal ways. One day he leaves home to enter the forest and is greeted by conspicuous No Hunting signs. Albert asks what this is all about, and he receives an answer: the land is now privately owned, and the owner wants Albert to stay out. Albert goes away, but the next morning he rises early and reenters the forest to hunt, without gaining permission. As he leaves around midday, police officers stop him. They arrest him for trespassing and take him to a police station.

Scene 2: Farmer Barbara has lived on bottomland for many years, growing food for home use and for the market. She grazes cattle and sheep on several pastures. One morning she rises to find the air filled with smoke and soot. Investigating, she learns that neighbors upwind are burning their fields. They have gone into the business of producing grass seed and need to burn their fields regularly to do so. As she investigates, she realizes that the grass burning not only sends smoke and soot into her house but significantly affects grassland birds that inhabit the region. When she makes inquiries at the state natural history survey, she is told that widespread burning is likely to stimulate many ecological changes. Insect species could rise in number, perhaps to pest levels, harming Barbara's crops. The grass growers are likely using chemicals to keep out weeds. These pesticides will also have ecological effects on plants, insects, birds, and rodents. But the truth, one scientist says, is that they really do not know what will happen as a result of the new grass seed business, given the ecological complexity of the bottomlands. Discouraged, Barbara drives home. On the way, she thinks about her long-held plan to divide her far pasture into building lots to sell for vacation homes. She fears her land will be worth much less if buyers must put up with smoke and soot and if their homes look out upon monocultural fields rather than natural-looking grasslands.

Scene 3 (further back in time): Harold is the head of an extended family clan, which tills its land using oxen. The land has been productive and yields a good surplus. One fall day armed men on horseback show up, carrying a strange banner. They are knights in the service of a nobleman named William, and they announce sternly that William has proclaimed himself owner of all he surveys. Henceforth, the knights assert, all land will be held subject to William's superior rights as lord and owner. All tillers of land will owe one-half of their produce to William, in recognition of his superior rights. The tillers will also owe 10 percent of their produce to the new church that William is constructing; a cousin of William's will be the local priest. As Harold contemplates the new situation, his eye on the horsemen and their weapons, he quickly calculates what this will mean. His entire farm surplus will be gone. He and his family will be reduced to bare subsistence. But what can he do about it, when William and his men hold the power?

What do these stories tell us about landownership and about the categories of lands commonly called public and private? What do they say about the way property ownership works, as an institution?

For starters, landownership in anything like the form we know it is a morally problematic institution in that it rests on the assertion of coercive power—that is, on the exercise of public or state power. When the new forest owner posts No Trespassing signs and then has hunter Albert arrested, he is obviously restricting Albert's liberties. Albert's freedom has diminished—not, ultimately, because of what the forest owner has done but because of the public power that the new owner wields. The law has vested this power in the forest owner, putting police and courts at his beck and call. Private ownership, in short, is all about the exercise of state power.

Now, an exercise of state power like this—physically taking Albert into custody—requires a good explanation to support it. It needs to be morally legitimate. It is not right to seize Albert and deprive him of liberty without good cause. Of course, we could say that Albert was arrested because he violated someone's property rights. But that is not a real answer; it is merely a paraphrase of the moral question. Why is it morally legitimate for one person to possess property rights that include this coercive power? Private property is the name we give to the power, not the justification for that power.

Our first scene, involving hunter Albert, shows how private property reduces the liberties of people who do not own land. Our second scene, involving farmer Barbara, shows how the exercise of rights by one property owner can conflict with the exercise of similar rights by another owner. Property rights are interdependent, and land-use conflicts arise regularly. A legal regime necessarily requires rules or processes to resolve the disputes that landowners regularly have. Somehow, the law needs to supply an answer here, deciding whether an intensive land use—in this case, burning grass stubble and applying pesticides—will or will not be permitted when it conflicts with the desires of landowning neighbors to be free of interference. These are not easy disputes, and we need to recognize that, in resolving them, there really is no pro–private property approach that we can take. Property rights lie on both sides of the dispute. Individual liberties lie on both sides of the dispute. The decision is not whether to protect private property; it is to decide what form of private property to protect.

Then there is our tale of Harold, and of marauding William, who takes over an entire country by fiat. Conquests like this happened, of course, and they could prove harsh for the people on the land. William's feudal property regime, instituted coercively, was essentially a way for him to control the people and to extract wealth from them. His property regime included a substantial element of theft. In this simple, rather ahistoric tale, almost everyone would agree that Harold has been mistreated, his farm produce seized, and his family reduced to poverty. But what would we say a few decades or generations later, when this new landholding system has become more familiar? Harold's descendents are peasants or serfs. Their lands are now controlled by their lord, securely in power. Perhaps the initial lord has sold his vast estates to some new lord, who now justifies the whole coercive arrangement on the ground that he has paid money for the lands—as he has. At some point, does the unfairness of it all disappear, or does the unfairness always remain? What if Harold's grandchildren rise up in revolt and refuse to pay rent on the ground that property is theft? Is it morally right for local constables to arrest them for withholding rent, or would the arrest be a continuation of William's original moral wrong?

Let me draw these various points together.

First, private property is primarily a form of power over people, not over land. To own land is to restrict what other people can do and sometimes to demand tribute from them. Property, in short, has a dark, coercive side. It expands the liberties and the powers of landowners but does so by necessarily restricting the liberties and economic options of other people.

Second, this power is necessarily a public power because it ultimately rests upon a landowner's ability to call upon police, courts, and even prisons to enforce his rights. We call property a form of private power, but it is misleading to do so. Ultimately, it is public power that private individuals are able to invoke.

Third, the exercise of power like this is morally problematic and therefore needs justification. Again, the justification we are talking about is a justification for restricting the liberty of people like hunter Albert, farmer Barbara, and yeoman Harold. Why is it legitimate to curtail the liberty of these people using state power? We need a good answer, and we cannot just point to the property rights as justification because it is the property rights themselves that need justifying.

These days, the kind of state power that supports private property is based on law, not military force. Private property exists to the extent it is authorized and supported by law. Maybe the law is morally legitimate; maybe it is not. But it is law that defines private rights. Take away the law, take away the public power, and the property rights no longer exist.[2]

This brings us to a final point, based on our tale of yeoman Harold. The power arrangement put into effect in that story, with William on top and with intermediate lords spread over the land, is recognizable as a feudal hierarchy. As ruler, William exercised what we now view as two distinct forms of power, although the distinction would have made little sense to William. We distinguish between the proprietary power that comes from property ownership and the public or sovereign power that comes from exercising governmental authority. When William the Conqueror took over England, proprietary and sovereign powers were fully mixed. His control over England's land brought him control over England's people. Were we to start in William's day nearly a thousand years ago and come forward to the present, we would see a long, uneven separation of these two sources of power. Slowly and erratically, the powers we label proprietary became separated from those that we view as sovereign or governmental.[3] Our tendency is to assume that this separation became total, that private property exists in a private realm and differs in kind from public or governmental power. But of course it does not. Private property rests upon public power and entails the exercise of that power.

As proprietary and sovereign powers split apart in England, most of the king's powers fell rather easily into one category or another, but some did not. Navigable waterways, for instance, resisted categorization.[4] The king owned the land beneath navigable waterways as a royal prerogative. But did he own this land as he might own a farm, or was it instead owned in some public or sovereign capacity? The same issue arose with respect to the king's rights over wildlife and beaches. If the king's powers over these resources were sovereign, then the public had claims to them and Parliament could regulate them. If the king's powers were proprietary, then the resources belonged to the king personally and could be kept for his exclusive personal use or sold.

Making matters more complex were the countless English common lands, for centuries subject to the claims of various people, usually residents of a local area.[5] Common lands were subject to use not by the public but by a defined group of people, often villagers. Rights to use the town commons were typically defined with precision. These rights were not private property, as we understand the term, because they entailed no exclusive control over any space. On the other hand, the common lands were not public lands because they were not open to the public generally.

As England marched to the present, economic forces pressed against medieval land-use patterns. Among the powerful changes were the waves of land enclosure that took place, mostly between the sixteenth and nineteenth centuries.[6] One type of enclosure occurred when a lord decided to get rid of the small farms on his land, consolidating and enclosing his fields and devoting them to sheep. To do this the lord evicted tenants, who often resisted bitterly. The other main type of enclosure took place on what had been common lands, subject to use rights by commoners.[7] These latter enclosures were authorized by acts of Parliament and ostensibly entailed payments to the evicted commoner (often in the form of land allotments rather than cash), but commoners resisted nonetheless because their economic and social lives were being upset. Enclosures that involved evicting tenants might strike us as legitimate exercises of a landowner's rights, but we need to remember that the feudal system was morally problematic. The tenants being evicted were the great-great-great grandchildren of yeoman Harold, who labored under a system that could be harsh and oppressive. In terms of the country's overall economy, the waves of enclosure might have made good sense; that is, economically the new land uses were often more efficient. But the unfairness and social dislocation nonetheless remained.

With this behind us, let us turn to the two categories of property that are familiar to us: public land and private land. They seem like different things, but how different are they? The points covered thus far help frame the answer.

Both public and private property are forms of power, meaning power that some people exercise over other people. Both are defined by law and indeed are creatures of law. Both are morally problematic in that they entail the coercive restriction of individual liberty. Both therefore need justification to remain legitimate.

When we turn to the laws that govern uses of private and public lands, we find that, like all other laws, such laws are rightly enacted only when lawmakers are attempting sincerely to foster the good of everyone, landowners and landless alike. Lawmakers are supposed to legislate for the common good, not for the benefit of any faction, and property laws are no different. Property is legitimate to the extent that it fosters the shared good.

This last point might seem surprising. We are accustomed to think about private property as an individual right of some sort, something that government is supposed to defend. After all, our constitutions contain protections for property, including protections against government invasion. So how can property be a product of legislative acts?

This is a vital question that needs, and has, an answer, though it would take time to review.[8] A key piece of the overall answer is that creating and protecting individual rights in land often furthers the common good. That is, individual private property can be a useful tool for fostering the good of people collectively. Moreover, private property works well only when owners enjoy reasonable stability in their rights. Lawmakers cannot just change the rules of ownership at will; if they do, the overall benefits of the institution decline. Individual property interests, then, need some protection. Nevertheless, it remains true that property is legitimate only when the governing laws promote the common good. Property becomes illegitimate, even oppressive, when owners are allowed to frustrate the common good, whether by harming other individuals or infringing public interests. Only secondarily is property an individual right.

The specific subject I've raised—the moral justification of property—is a complicated one. Indeed, the whole institution of private property is complex. Private property is also fascinating to study, in terms of its history, its varied manifestations, and the ways diverse peoples have talked about it. The powers and obligations of landownership have varied greatly across time and place. The ideas we embrace in the United States today are the product of our time.

Back to the public-private divide. So far, I have highlighted essential ways that public and private lands are the same: rights to control these lands are forms of power; these powers derive from law; the laws are morally problematic; and the laws are justified only when they foster the common good. The two categories are thus similar, and the divide between them is narrow. It is simply not the case that private rights exist apart from law, or foster private interests apart from the public good, or exist as a form of private power that is independent of public power.

How then do the public and the private differ, because surely they do?

To get at their differences, we can return to our opening scene, with our people entering their new land. As the people gaze upon the landscape, thinking about how they'll inhabit the landscape, they confront three questions.

First, how are they going to use these lands to foster their collective good?

Second, who is going to use which lands?

And third, who gets to make the decisions?

These are the vital questions, both for the first people who enter a place and for each generation that follows. By keeping the questions front and center, we can get at the differences between public and private lands.

The biggest difference between private and public lands has to do with management power over the land. Who gets to decide land uses? Decisions about public lands are mostly made by public decision-makers, but not completely so. Public decision-makers are often influenced by private parties who want to use the lands. Indeed, private involvement in public-lands processes is extensive, too extensive, some people say. When we turn to private lands, the equation is flipped but again is not one-sided. Private owners have greater say in land-use decisions, but lawmakers commonly play important roles; again, too important, some people say. In many settings, private lands are also subject to limits imposed by other private citizens—by a homeowners' association, for instance. In both cases, then, public and private influences intermingle. So varied is this intermingling that we do not really have two categories of lands. We have a continuum, with some lands more subject to public control and some lands more subject to private control. Control of either type is always a matter of degree.

On the question of how the land is used, we also see a continuum or mixture of uses rather than two distinct types. On public lands we have nature preserves, intensively used parks, grazing, logging, mining, office buildings, stores, and so on. On private lands we find pretty much the same, less in the way of nature reserves and more in the form of intensive land uses, particularly residential uses. Without maps or signs, though, it is often hard to tell public from private.

We get even greater overlap when we consider who actually uses the land. Private lands are used by private actors almost exclusively. But activities on public lands also involve private actors; indeed, private parties are the primary users of public lands. Logging, grazing, timber harvesting, mining, recreation—all are undertaken on public lands by private parties, usually at some private initiative. So again, in terms of land use, the differences between public and private actors are ones of degree. The public and the private overlap.

Consider, for a moment, the typical residential subdivision lot, a familiar form of private property. The owner's use of the lot is probably subject to severe limits as a result of restrictive covenants, enforceable by neighbors. Permanent easements might allow public utilities or even private entities to enter this residential lot and use it for specified purposes. Zoning laws could limit activities or even prescribe affirmative duties, such as shoveling snow, maintaining fences, and keeping weeds trimmed. When we get down to it, the owner of this lot might really have only a single, narrowly defined use right in the land—a specific right to use the land for a single-family home.

Compare this carefully prescribed residential-use right with a similar right to use public land, such as a Bureau of Land Management grazing permit or a federal oil and gas lease. Here, too, we have a private property right, and it is carefully tailored by law. So how different is the grazing permit from the homeowner's use right? There are differences, to be sure, yet both are specifically tailored use rights, both are largely defined by law, and both are crafted, one hopes, so that the private activities promote the common good.

The categories that we know as public and private land have not always been around, certainly not in anything like the way we commonly think of them. For decades after the United States arose, it was assumed that pretty much all public domain land would pass into private hands. It was not even particularly clear that the government's rights as landowner were more extensive than any other landowner's.[9] As for private land, there was a long period during which rural areas were mostly an open commons, where people could roam, hunt, forage, graze livestock, and collect firewood without the landowner's permission.[10] In fact, the landowner's right to exclude outsiders was largely limited to areas that were fenced or cultivated. In most rural areas, the woods were one great commons.[11] It made little difference what land was public and what land was private.

During the nineteenth century, private property rights in the United States changed considerably with respect to the powers that landowners possessed.[12] A landowner's right to exclude outsiders expanded, particularly on unenclosed rural lands. A landowner's rights to use the land intensively also expanded, even when the new, industrial land uses harmed surrounding lands. By the mid- to late nineteenth century, the idea emerged, really for the first time, that landowners could largely do whatever they wanted, so long as they didn't cause visible, substantial harm to other people. This was a distinctly pro-industrial vision of private property, quite different from the agrarian approach to property ownership that prevailed a century earlier and that limited the ways landowners could use their lands. Necessarily, this new pro-industry approach to ownership meant that sensitive land uses were no longer well protected against interference by noisy, industrial neighbors. Over the course of the nineteenth century, the landowner's right to use the land intensively expanded, while the landowner's right to halt interferences contracted.

By 1900, the law empowered private landowners to do many

things that were widely deemed unwise or destructive. Not surprisingly, private property came under attack. Criticism came not only from the new generation of conservationists concerned about overcut forests and degraded waterways[13] but also from other progressive reformers, particularly in cities.[14] As many people saw matters, property law gave landowners too much power to act selfishly. Mining companies and meatpackers polluted waterways; rising industries degraded residential areas. Fertile soil washed away while fires spread through cutover forests. Things needed to change. Private property had become a serious problem.[15]

One tool reformers used to address the ills of private ownership was land-use regulation, a form of public control with a long history in America, dating back to early colonial settlement.[16] A new, more comprehensive approach to land-use control arose with the coming of modern zoning laws, which divided urban areas into zones and prescribed the uses permissible in each.[17] Regulation, though, was not the only response to private property's ills. Also welling up was a call for the nation to hold on to its public lands and to manage these lands for the long-term public good. If private owners were not going to use their lands sensibly—if they were going to cut down the North Woods and plow up the plains to create a dust bowl—then the public would have to look to public lands for amenities such as healthy forests, unpolluted rivers, and pleasing recreational spaces. And so the call went out: Hold on to public lands. Halt the era of land disposal. It was a momentous decision—taken not at once but over several decades—for a nation that had not intended to stay in the landowning business.[18]

The point here is an important one, worthy of emphasis. *We have retained expansive public lands in the West in large part because of the perceived failings of private property.*[19] When private landowners can degrade their lands and get away with it—even though private property is supposed to support the common good—then it is understandable that people will want more public land, and that they will want their public lands protected from being used in the same ways that private lands are used. Public land was the remedy for private irresponsibility.

This lesson that reformers learned early in the twentieth century was not the only lesson they could have learned. But it was the obvious lesson, given the then-prevailing ideas about the powers that private landowners possessed. The assumption of the day was that private owners could degrade their lands if they chose. They could strip their trees, plow fragile soil, and dig up minerals, all with little regard for the land's long-term health. The dark side of this individual liberty appeared unmistakable in the Dust Bowl decade of the 1930s, when homesteaders plowed land that should have remained in grass.[20] We call the Dust Bowl a natural disaster, but the problem was caused by people, not nature, as thoughtful reformers at the time could readily see. And the solution, reformers said, was to halt further land disposition and to create a federal grazing service. Whenever it was possible, the government would also buy back degraded private wheat fields and return them to publicly managed pasture.[21]

In terms of land-use errors, reformers in the 1930s were assessing the situation accurately: semiarid land should be grazed, not plowed. Where the reformers fell short in their understanding was in assuming that grazing could be assured only on lands that the public owned. They did not understand that land could be turned over to private hands subject to a legal requirement that the land not be plowed. Private property is a more flexible arrangement than the reformers understood. It need not give landowners freedom to misuse what they own. Public landownership was not the only remedy for misuses of private land.

The public-private divide as an intellectual framework, as a way of thinking about our current land-use regime, is distinctly unhelpful today. It implies that some lands can be used solely for an owner's benefit while others are used for the good of everyone. Yet that division makes little sense. The public has a legitimate interest in how all lands are used. No land use takes place in isolation. As for public lands, many are needed to serve distinctly public purposes, but most are not. Or rather, most publicly owned lands would not be needed to serve public activities if we could be confident that, when the land was placed into private hands, private uses would comport with the common good.

We find ourselves today, I think, burdened with several lousy ideas that we would do well to alter or discard.

The most pressing of these lousy ideas is that private property includes the right to use the land any way an owner wants, without regard for public implications. This is not an accurate statement of law or history, nor is it remotely good public policy.

A second lousy idea in need of change is that the only way to promote healthy lands is to keep them in public hands. Neither is this true, however understandable the idea was when it arose about a century ago.

A third lousy idea is that we can sensibly define the property rights a landowner possesses without taking nature into account. The idea here is that property rights in a tract of land—in the hypothetical Blackacre or Greenacre, as law students would label it—can be defined in the abstract, without regard for the land's natural features. Land parcels in fact differ greatly, and the differences in their natural features affect how we can safely use them. In defining land-use rights, we need to take nature into account. And we are doing so, albeit slowly and in ways that arouse controversy. The private rights of landowners are now much different in wetlands and floodplains, on barrier islands and beaches, on sloping hills subject to erosion, in forests and critical wildlife habitat, and along riparian corridors. It is easy to view our complex legal regime today as simply a collection of isolated laws and regulations, federal, state, and local, each aimed at some specific environmental problem or land-use concern. But the laws and regulations collectively do form a pattern. They reveal a distinct trend of looking to nature itself to help us decide not just how to use lands but also how to define the legal rights that landowners possess.[22]

In my view, we have too much public land today. We also have misguided ideas about what private property entails. And the two problems are linked. We have one problem because we have the other, and we cannot deal with one unless we deal with both.

So how might we deal with these problems? What would things look like if we replaced our lousy land-use ideas with more sensible ones, with ideas based not upon a presumed chasm between public and private but instead on a recognized need to combine public and private nearly everywhere, on all lands?

The virtue of private ownership is that it designates particular people as land stewards, charged with looking after the land and putting it to good use. Private ownership can protect privacy, provide incentives for economic enterprise, and add ballast to civil states. Public ownership, on the other side, is better able to consider the long term and can assess land uses in broader spatial contexts. Government can resist market pressures to misuse land, and it can manage lands to provide an array of public goods that make little economic sense for individual owners. Of course, both forms of ownership can and do fall short of the ideal. Private owners are often not good stewards: their perspectives are too short, they ignore ecological ripple effects, and their isolated decisions can produce chaotic land-use patterns. Government agencies, on their side, are buffeted by political winds and have trouble saying no to powerful groups. Their decisions can be painfully slow and inefficient.

The main challenge we face today in attempting to live well on land—attempting to succeed at the "oldest task in human history"—is coming up with better ways of combining public and private on the same piece of land. The public has a legitimate interest in the way all land is used, private land very much included. In the case of private land—as our current land-use squabbles illustrate—we are having trouble finding good ways to protect that public interest without undercutting the vital benefits we all get from a scheme of widespread private ownership. How do we protect the public's interest while retaining the important benefits we get from a private property system? That is the question. We need better answers.

This need to protect the public's interest in private land is particularly vital because it goes to the heart of private property's legitimacy. As noted, private property in land is not morally legitimate when it allows owners to harm the public good. After all, why should we deploy our police and courts to support private action that harms the community? That simply is not right.

No law, of course, can ever be so precise as to prescribe the exact ways that land should be used. Laws are crude tools, and they can do little more than restrict the most harmful practices. To get truly good land use, landowners have to want to conserve. They need to know the nuts and bolts of sound, conservative land use as applied to their own lands. That said, though, there is a lot of room to improve the institutional context of private land use so as to increase the influence of public values in private land use decision making. And the place to begin, in asserting this public interest, is with the basic rights that landowners possess. If plowing a hillside can lead to degradation, harming the public as well as the landowner, then why should the landowner have the legal right to plow? Why should that be a component of the landowner's bundle of entitlements created and supported by public action?

New laws could better protect the public interest. We could call these laws property laws, or we could call them regulations; it makes little difference. We could also protect the public interest using mechanisms that are less obviously public. Examples here include restrictive covenants, rules imposed by homeowners' associations, and restrictions that come through resource-management cooperatives. There is also the public involvement in private decisions that takes the form of economic incentives to use land in publicly good ways, whether funded by taxpayers or private donors.

These familiar ways of combining public and private, though, need to be understood as merely illustrations of what is possible, perhaps as precursors of more effective methods that await our courage and imagination.

Consider, for instance, a grazing arrangement, the Tilbuster Commons, that has been put together in eastern Australia.[23] Under it, private landowners lease their private lands to a collectively managed grazing cooperative. Their combined lands are worked in concert—like open-field farms of centuries ago—with their animal herds mingled. By working jointly the grazers can employ a larger spatial perspective in their land management, thereby reducing one of the main defects of traditional private ownership. Here in the United States, we have similar examples of cooperative land management, such as the pooling and unitization schemes that govern oil fields[24] and water-management schemes orchestrated by water conservancy districts.[25] Safe-harbor and candidate-conservation agreements under the Endangered Species Act offer useful precedents,[26] as do federal agencies' experiences managing grazing, timber harvesting, and mining on federal lands. The new Forest Service program involving stewardship contracts illustrates a willingness to try new public-private land management forms and could prove a step in the right direction.[27] Across the West, there is talk about connecting private and public grazing lands in ways that view them as integrated management units. Again, though, these are just hints of what is possible when we stop thinking about land as either public or private and instead look for new ways to combine public and private on all lands.

For the vast majority of lands, where we need to head (and are heading, haltingly) is toward blended landscapes, in which private actors possess use rights that are loosely tailored to protect the public interest. These use rights are forms of private property, but they bear little resemblance to the industrial, ownership-as-absolute-dominion ideal of private property that arose in the nineteenth century. Tailored use rights have existed for years on public lands. On public lands, we are likely to see an expansion of these use rights so that private holders can plan over longer time periods and can take broader responsibility for the land, subject to duties to take good care of it. These private use rights on public lands will not typically be exclusive; the land might remain open to public recreational use, for instance, and a holder of timber or grazing rights might need to defer to someone else who holds mining rights. But public recreational rights might be more limited than today; they might be limited to public hiking on defined trails, without ATVs, snowmobiles, or even mountain bikes. In addition, the holder of a private use right might have the power and duty to halt destructive trespasses.

Tailored use rights could look pretty much the same when they exist on private land. For a look into the future of private landownership, we might consider the case of timber harvesting in a state that is aggressive in regulating forestry to protect nature.[28] In such a state, a forest owner can be restricted by law from harvesting trees along waterways or near residential areas. A state forestry practices statute could require the owner to preserve the diversity of tree species and ages while limiting harvesting methods and imposing duties to replant. Perhaps the forest owner has already sold hunting rights to a local hunting club, and perhaps an old railroad right-of-way or mining road is used as a public hiking trail. Perhaps there is even a conservation easement on the land. When we put all these elements together, our hypothetical private forest already might look a lot like a public forest, in terms of the legal rights that the timber company holds and the ways multiple uses of it are mixed.

As we look ahead, we are likely to see new ways in which the public interest in land is identified and protected. We will rely less on distant governments and instead make greater use of novel, collective-management arrangements that are closer to the land. We are also likely to have the public interest refined and promoted by multiple levels of government that pay attention to differing spatial scales. Perhaps we will even see more arrangements that involve collaboration, cooperation, and adaptive management, undertaken by groups of people whose roles blend the public and private, groups, like today's homeowners' association, that are essentially private in operation but recognized by law and subject to legal constraint.

I predict for the future a marked reduction in public lands as we now know them. This will be good news to some. But this should happen only if we also experience an equally marked reduction in our passionate embrace of outdated ideas about private landownership. It should happen only if we have an even greater reduction in private lands as we now know them. The better we protect the public interest in private lands, the less need we will have for overtly public land. We do not need a shift of land from one type of ownership to the other. We need instead an end to the categories themselves. We need to craft new, intermediate forms of land management, and then shift lands from both sides into the center.

In the end, public land and private land are really not all that different when we look at them closely, in terms of who uses them, how they are used, how use rights are defined, who makes decisions, the need for moral justification for the governing laws, and the ultimate duty to limit uses so as to foster the common good. We do not have two categories of land. We have instead a wide-ranging continuum of public and private interests on the same lands. And that is the way it should be. Indeed, we need even more and better-crafted blends of public and private interests, which do a better job taking advantage of the best elements of both private and public ownership.

To craft better land-management arrangements, we need first largely to cleanse our minds of these two categories. The categorization has itself caused distinct problems in our thinking about land and land uses. Problems in our thinking in turn have led to needless conflict and some bad land-use decisions. We have only one category: land. The public has a legitimate interest in how all land is used. By the same token, the principal users of lands are almost always private users, and there are good reasons that private use rights in land might be defined as forms of private property.

If we want, then, a simple image of land, it should be this: The land is owned ultimately by the sovereign people collectively, the demos, and managed for the common good. But private parties have use rights in this land. We thus have two items to discuss for nearly all land: what private use rights should look like and what mechanisms we should develop to ensure that these use rights and the management of lands generally promote the common good. Use rights and collective management regimes: those are our topics and our challenges, for all lands. The possibilities are countless; the room for improvement is vast. We need to get to work.

Notes

1. Aldo Leopold, "Engineering and Conservation," 1938, in *The River of the Mother of God and Other Essays*, ed. Susan L. Flader and J. Baird Callicott (Madison: University of Wisconsin Press, 1991), 254.

2. This point is firmly established in American jurisprudence. See, for example, *Fox River Paper Co. v. Railroad Com'n of Wisconsin*, 274 U.S. 651, 657 (1927) (the Fourteenth Amendment "affords no protection to supposed rights of property which the state courts determine to be nonexistent").

3. Dale D. Goble, "Three Cases/Four Tales: Commons, Capture, the Public Trust, and Property in Land," *Environmental Law* 35 (2005): 807, 820–22.

4. Ibid., 824–29.

5. E. P. Thompson, *Customs in Common: Studies in Traditional Popular Culture* (New York: New Press, 1993), 97–184.

6. Ibid.; Mark Overton, *Agricultural Revolution in England: The Transformation of the Agrarian Economy, 1500–1850* (New York: Cambridge University Press, 1996), 147–67; G. E. Mingay, *Land and Society in England, 1750–1980* (New York: Longman, 1994), 35–46.

7. The term "enclosure" included various other ways in which open landscapes were transformed into discrete fields, surrounded by high fences (or enclosures). In medieval open-field farming systems, individual landowners (who held their lands subject to a lord, of course) often possessed individual rights in various small, scattered tracts of land, intermingled with similar tracts of land owned by their neighbors. Much of the farming was done communally, or at least according to work schedules agreed upon by the village members. One form of enclosure took place when the small farmers exchanged lands with one another, allowing each to end up with a single, contiguous landholding, which could then be enclosed and used separately. Many times, one owner would buy out another, just as use rights in common lands were bought out by the owner trying to enclose the commons.

8. I offer an answer in Eric T. Freyfogle, *The Land We Share: Private Property and the Common Good* (Washington, DC: Island Press, 2003), 101–32.

9. George Cameron Coggins, Charles F. Wilkinson, and John D. Leshy, *Federal Public Land and Resources Law*, 5th ed. (New York: Foundation Press, 2002), 49–51, 182–84.

10. Only small pieces of this important story have been told. Glimpses are offered in Richard W. Judd, *Common Lands, Common People: The Origins of Conservation in Northern New England* (Cambridge, MA: Harvard University Press, 1997), 28–56; Stuart A. Marks, *Southern Hunting in Black and White: Nature, History, and Ritual in a Carolina Community* (Princeton, NJ: Princeton University Press, 1991), 32 ("Open land, which encompassed most of the land in the South, was considered [in the antebellum era] as common property for hunting, fishing, and grazing"); and Steven Hahn, "Hunting, Fishing, and Foraging: Common Rights and Class Relations in the Postbellum South," *Radical History Review* 26 (1982): 37. Public rights to use the countryside were most firmly established in the South. For instance, one 1860 case involved a horse owner who sought to recover damages for the death of his horse by a train. The court summarily dismissed the railroad's claim that it should escape liability because the horse was trespassing on its tracks: "Such law

as this would require a revolution in our people's habits of thought and action. A man could not walk across his neighbor's unenclosed land, nor allow his horse, or his hog, or his cow to range in the woods nor to graze on the old fields, or the 'wire grass,' without subjecting himself to damages for a trespass. Our whole people, with their present habits, would be converted into a set of trespassers. We do not think that such is the Law." *Macon & Western Railroad Co. v. Lester*, 30 Ga. 911 (1860).

11. See *M'Conico v. Singleton*, 9 S.C.L. (2 Mill) 244, 246 (S.C. 1818).

12. See Freyfogle, *Land We Share*, 65–84 and sources cited.

13. Judd, *Common Lands, Common People*; Samuel P. Hays, *Conservation and the Gospel of Efficiency: The Progressive Conservation Movement, 1890–1920* (Cambridge, MA: Harvard University Press, 1959).

14. Robert Gottlieb, *Forcing the Spring: The Transformation of the American Environmental Movement* (Washington, DC: Island Press, 1993), 47–75; Donald Worster, ed., *American Environmentalism: The Formative Period, 1860–1915* (New York: Wiley, 1973), 111–82.

15. Donald Worster, "Private, Public, and Personal: Americans and the Land," in *The Wealth of Nature: Environmental History and the Ecological Imagination* (New York: Oxford University Press, 1993), 95, 103 ("The conservation movement emerged out of discontent with an intensely private approach to land ownership and rights. It has been an effort to define and assert broader communitarian values, some idea of a public interest transcending the wants and desires of a strictly individualistic calculus").

16. The early history of land-use regulation is surveyed in John F. Hart, "Colonial Land-Use Law and Its Significance for Modern Takings Doctrine," *Harvard Law Review* 109 (1996): 1252.

17. Rutherford H. Platt, *Land Use and Society: Geography, Law, and Public Policy* (Washington, DC: Island Press, 1996), 269–84.

18. The story is recounted in E. Louise Peffer, *The Closing of the Public Domain: Disposal and Reservation Policies, 1900–50* (Palo Alto, CA: Stanford University Press, 1951).

19. The idea is expressed, although not quite in these words, in Worster, "Private, Public, and Personal," 103–4.

20. Donald Worster, *Dust Bowl: The Southern Plains in the 1930s* (New York: Oxford University Press, 1979), 87–97.

21. U.S. Department of Agriculture, *The Land Utilization Program, 1934 to 1964: Origin, Development, and Present Status*, Agricultural Economic Report no. 85 (Washington, DC: Economic Research Service, USDA, 1965), 1–40; William D. Rowley, *M. L. Wilson and the Campaign for the Domestic Allotment* (Lincoln: University of Nebraska Press, 1970), 133–34.

22. I consider this and other major trends having to do with ownership rights in nature in Eric T. Freyfogle, "Community and the Market in Modern American Property Law," in *Land, Property, and the Environment*, ed. John F. Richards (Oakland, CA: ICS Press, 2002), 395–401.

23. Sima Williamson, David Brunckhorst, and Gerard Kelly, *Reinventing the Common: Cross-Boundary Farming for a Sustainable Future* (Annandale, NSW: Federation Press, 2003), 22–30.

24. Owen L. Anderson et al., *Hemingway Oil and Gas Law and Taxation*, 4th ed. (St. Paul, MN: Thomson/West, 2004), 386–99.

25. See Barton H. Thompson, "Institutional Perspectives on Water Policy and Markets," *California Law Review* 81 (1993): 671.

26. These developments and others under the Endangered Species Act are surveyed in J. B. Ruhl, "Endangered Species Act Innovations in the Post-Babbittonian Era—Are There Any?" *Duke Environmental Law and Policy Forum* 14 (2004): 419.

27. U.S. Department of Agriculture Forest Service, *Stewardship End Result Contracting Policy*, 69 Fed. Reg. 4107 (January 28, 2004) (notice of issuance of agency interim directive governing stewardship contracting).

28. A critical look at California's forestry regime is offered in Thomas N. Lippe and Kathy Bailey, "Regulation of Logging on Private Land in California under Governor Gray Davis," *Golden Gate Law Review* 31 (2001): 351.

6

BACK TOWARD COMMUNITY

The institution of private property is one of the chief mechanisms through which a society interacts with the natural world. Property law creates a framework for managing and using nature. It explains who gets to do what, and where. When a landscape is divided into private parcels, it is not the land that is fragmented; nature remains an integrated whole. What is fragmented is the legal power to make decisions over land. The law prescribes how individuals can acquire managerial powers over particular segments of this natural whole.

One question that presses itself upon American law and culture today is whether the rules that prescribe property rights ought to be influenced, more than they are, by nature itself. Should the legal rights a landowner possesses vary based upon the physical features of the land? Should the right to reshape or develop land, for instance, depend upon terrain features or the plants and animals that live there? Much of what we know about nature's functioning is assembled within the scientific discipline of ecology. We can therefore rephrase this question: should our ecological understandings of land help shape our legal framework of private ownership, in terms of what can be owned and the rights and responsibilities that landowners possess? When it comes to land, property law and ecology are kindred disciplines. Ecology explains how activities on one land parcel can affect landowners and land uses elsewhere.

Property law then evaluates these spillover effects, deciding whether activities that cause the spillover effects are lawful.

To answer this elemental question about landowner rights and nature, it is useful to bring ethics into the mix of relevant considerations, for reasons that will become evident. When we do this—when we consider private property, ecology, and ethics together—we are better able to spot patterns of overall cultural development. A pattern of conflict and change that is obscure in one of these three fields can become clearer when we see the same pattern in another. A development that is seemingly contained within one field can appear different when we realize that it is a piece of a larger cultural change, driven by forces that transcend and encompass the field itself.

These three, related fields of knowledge are similar in that each is characterized by an inner tension or duality. And it is the same tension, the same duality or fault line, that runs through much of American culture. We can describe it, for the moment, this way: On one side is the tendency toward abstract reasoning and individualism or atomism. On the other side is the tendency toward particularity, context, and community. The first side emphasizes the parts and accords them independent status. The other side points toward the connections and dependencies among the parts and highlights the importance of well-functioning interactions. This tension or dichotomy, once we see it at work, provides a useful framework for understanding the trajectories of these related fields. It helps us to see how the fields are buffeted by larger cultural forces and to identify the cultural changes that might be necessary for society to enhance the health of its lands and communities. Looking ahead we face a choice: should we move in the direction of greater abstraction and individualism or instead in the direction of greater attention to connections and dependencies? For those who desire a culture well rooted in land, the answer is rather clear.

We can begin the story in Cambridge, Massachusetts, in 1870. Harvard College at the time was in poor shape, and Harvard Law School was even worse. Higher education had fallen on hard times in the midst of one of America's periodic waves of anti-intellectualism. Enter Charles William Eliot, who was hired by the leaders of Harvard College with the charge of shaking up the place to give it respectability and rigor. Eliot was a mathematician and chemist by training, and he believed he knew just what Harvard needed. The home of intellectual rigor, he asserted, was in the hard sciences, particularly the physical sciences. The way to straighten up Harvard, accordingly, was to infuse its various programs with the logic and methods of inductive science.[1]

To take over the law school, Eliot brought in Christopher Columbus Langdell. Bold and decisive, he was just the man for the job. Legal study, Langdell decided, should be reorganized so that it mimicked what went on in the chemistry laboratory. Just as chemists entered their labs, gathered data, and identified the basic principles about how elements and molecules worked, so too law students should go into their laboratories, collect their data, and distill the basic principles that similarly guided the law. And where was that laboratory? It was the law library. And what were the data? They were the reported decisions of appellate courts. What legal study should entail, Langell decided, was the careful parsing of judicial opinions. From them, a student could distill the operative principles of law, which he then could apply mechanically in new factual settings. "The library is the proper workshop of professors and students alike," Langdell explained. "It is to us all that the laboratories of the university are to the chemists and physicists."[2] The era of dry, textbook lectures had come to an end, Langdell hoped, and the inductive case method of law study had begun.

Langdell's scheme was wonderfully simple even as it helped make legal study academically respectable. Langdell rejected the idea that law practice was merely a trade that a person could learn by apprenticing. Law was a serious field of study, he said, worthy of room in top-level colleges. To learn the law, a student need not and should not engage with the messy outside world. There was no need to pay attention to actual people, struggle with ethical enigmas, or know anything about nature. Indeed, a student never even had to leave the law library. Not the vicissitudes of life but rigorous logic should infuse the law, supplying order and guiding future court rulings.

So radical was Langdell's approach that he had to find a new breed of law teacher to implement it, teachers who would not confuse law with life. "What qualifies a person . . . to teach law," Langdell announced boldly, "is not experience in the work of a lawyer's office, not experience in dealing with men, not experience in the trial or argument of cases, not experience, in short, in using law, but experience in learning law."[3] Dean Ames, Langdell's handpicked successor, was equally emphatic that the law school's intellectual boundaries should remain firm: "We are unanimously opposed," Ames pronounced, "to the teaching of anything but pure law in our department. . . . We think that no one but a lawyer, teaching law, should be a member of the Law Faculty."[4]

Langdell's approach fostered a highly abstract understanding of law. From prior judicial decisions, it distilled legal rules that were stripped of nearly all factual details. In the law of contracts—Langdell's personal field of scholarship—contracting parties were treated at law as equals, regardless of personal wealth, power, age, or experience. The law overlooked the reality that flesh-and-blood people differed greatly in their understanding, sophistication, and bargaining power. All contracts were legally enforceable, without concern for whether the price was fair, whether the contracting parties understood what they were doing, and whether the contract benefited the persons affected by it. Older legal principles upheld a contract only if they were based on a "fair price" and if the parties understood their agreement in the same way. In the new era of abstraction, these messy particulars faded away. After all, who was to say whether a price was fair or what a person thought when signing a document. Logic alone could not answer these questions. By ignoring them, the law effectively transformed real-life people into lifeless abstractions: party A and party B.

In the case of property law, the rights of landownership were expressed in ways that similarly paid little attention to distinguishing traits. Land was land, and a parcel's physical features and surroundings were legally irrelevant. Indeed, law students did not even talk about real physical places and real owners. They talked instead about hypothetical tracts of land—Blackacre and Greenacre. Blackacre's owner possessed rights that were defined as abstractly as the rights of contracting parties. Blackacre's natural features (its soils, slope, plants, animals, climate, water regime) as well as its social setting (who the neighbors were, what they did, what the surrounding neighborhood was like) were of little consequence. In the view of abstract theorists, landowners' rights ought to be essentially the same for all lands everywhere.

The owner of this hypothetical Blackacre, law students learned, was subject to the rule that she could not engage in activities that harmed others. Do-no-harm was an established legal rule, expressed in a venerable Latin phrase. But though the phrasing of the rule had remained stable for generations, the practical definition of "harm" certainly had not. In the real world, where judges and juries decided cases, "harm" was a term that shifted in meaning over time, right along with the rest of culture. When the nineteenth century began, the do-no-harm rule imposed a substantial limit on the power of landowners to use their lands intensively in terms of noise, pollution, vibrations, stench, and the like. A landowner could not harm neighbors or the surrounding community in any noticeable way. As the century wore on, however, the term "harm" took on more limited meanings. It imposed less of a restraint on what landowners could do. The rules of property owning were on the move.[5]

The problem that arose as the nineteenth century unfolded was that a strict application of the do-no-harm rule of land ownership created big obstacles for industries, railroads, and mines. Industrialism meant using lands intensively, with the frequent result that industrial landowners interfered with the "quiet enjoyment" of landowning neighbors. Economically, it was difficult or impossible for industries to operate if they had to pay for every harm they caused. The problem was even worse when a neighboring landowner who suffered harm could get a court order to shut the industry down. Most courts had no trouble understanding the high stakes, economic and legal, involved in these conflicts. As judges saw matters, property law simply had to change. It had to make room on the landscape for these new, economically valuable industries. In many states and to varying degrees, the legal result was a redefinition of the concept of land-use harm. Courts redefined "harm" so that only the most severe and unreasonable harms violated the rule. As for lesser harms, the law ignored them. Landowners who suffered lesser harms would simply have to grin and bear it. The result of this legal change was that landowners gained greater powers to engage in intensive activities. At the same time, they suffered a corresponding reduction in their legal rights to complain when neighbors disturbed their activities.

Christopher Columbus Langdell did not discover the idea of law as science. And the big shift in nineteenth-century property law, to allow more intensive land uses to take place, was largely complete when he arrived at Harvard. What Langdell did was construct a logical framework to explain this pro-industry vision of owning. He gave to the abstract Blackacre, detached from real life, the imprimatur of the physical sciences.

This Victorian-era vision of private ownership did not arise without complaint. Far from it. Many people, landowners included, were injured or even killed by the new industrial activities—from the polluted air and water, the noise and stench, the artificial floods and fires, the disappearing fish and game, or simply the ugliness of the new landscape. As many people saw things—people who cared about health, beauty, and settled communities more than they did about rapid economic growth—the reasonableness of a particular land use ought to depend on where it was taking place. An intensive land use might be fine in one location but not fine in another. Context, they stressed, was critical. The rights of ownership in a land parcel ought to depend upon its natural and social settings. The idea was hardly shocking. Indeed, the common law had once embraced a version of the idea. But Langdell's scientific construct had little place for it.

Langdell's vision of property, so warmly embraced by industrialists, developers, and political libertarians, was of course an atomistic one. Langdell saw the landscape divided into separate, independent pieces of land that owners could use without much regard for context. The individual was all; the surrounding community counted for little. Just as one oxygen atom had the same traits as all others, so too the rights of one landowner should be the same as those of the next.

To see how stark Langdell's vision really was, we can compare it with the property regime that existed hundreds of years earlier, in England. In feudal times, land-use rights were situated in a hierarchical social order. No one save the king held land outright; for everyone else there was always a higher lord to whom fees and loyalty were due. Particularly in the generations immediately after the Norman Conquest, the aims of property law had little or nothing to do with individual rights or with the promotion of individual initiative. Property upheld a complex, tight-knit social world in which each person had a place and was expected to keep to that place. Society was an organic whole, military strength was a critical need, and property law kept it all together.

These feudal beginnings are usefully recalled because they provide a place to begin tracing the trajectory of Anglo-American property law since the late eleventh century. Property rights in William the Conqueror's day promoted the common good as he and his vassals understood it. The community was essentially all, with William at the head and society arranged along military lines. The individual as such (rich ones aside) counted for little. At the lower levels of the hierarchy, individuals were constrained by manorial customs and local rules. Because of collective land-management processes, tillers of land often had little choice in their land-use decisions.

From Norman England to the days of Langdell, we can draw a crude line charting the path that private property took. From an institution in which context and community were influential, it shifted to one in which the isolated individual held powerful rights, defined with little regard for nature, society, and communal needs. A similar shift took place in thinking about the market—from a beginning point where the market was relatively unimportant to a place where market thinking about land held dominance. Individually the steps along the way were small ones, but in time they ushered in massive change.

This wide-ranging shift, of course, did not come about by happenstance. Blood was spilled. Kings were confronted. Economic interests pushed and conspired, as winners rose and losers fell. John Locke's labor theory of property, so popular in the seventeenth and eighteenth centuries, needs to be understood in the context of this shift. Locke claimed that an individual became owner of a thing by mixing labor with it and adding value. Property rights, that is, arose through individual initiative, not through any royal or communal decision to create the property rights and allocate them. Locke's state-of-nature story was historically fanciful, but it struck a responsive chord. Locke also claimed, counterfactually, that property was essentially a timeless, intangible idea, in that property largely meant the same thing to everyone at all times (otherwise how could property arise in a state of nature?). Locke's theory had the effect of situating private property outside society; property was a natural right, not a social convention. This was a good result, in Locke's view, because it supported and extended the ongoing destruction of feudalism. The theory promoted Locke's obvious aim, which was to move property further along in its journey from Norman England to a world of liberal individualism.

If we were to carry forward this story, from Langdell's day through the twentieth century, we would find that individualism and abstraction were extended too far, at least in the view of most lawmakers. Intensive landowners were doing too much harm to their surroundings. The view of land as market commodity was proving disruptive for landscapes, communities, and individuals. New laws were needed to curtail the most disruptive activities and to harmonize land uses.[6]

Governmental land-use controls had long been around in America. Indeed, regulatory limits were moderately common in colonial times, particularly in cities. When landowners complained about these land-use limits, courts consistently upheld governmental powers to regulate. Even in the Victorian era—the age of the robber barons—it remained clear that governments had the legal power to halt the most damaging activities. The obstacle that reformers faced as they pressed for change was more intellectual and cultural, than legal. When property was viewed as an abstract bundle of rights, defined out of any context, then land-use rules could easily appear suspect, to ordinary people as well as lawyers. Land-use regulations that took context into account seemed to cut into the abstract private rights of owners. Land-use limits also seemed to clash with the nation's conquering urge to tame the land. America was a society based on changing the natural order. To limit change clashed with this pioneering culture.

Even with these cultural conflicts, however, communities did take steps to curtail the worst effects of the industrial age. Some of the new laws addressed specific land-use problems—the polluting slaughterhouses and tanneries, for instance, which were killing fish and rendering waters undrinkable. Other new laws were more far-reaching, particularly the new zoning rules that imposed land-use limits over whole cities. The effect of these laws was to force industrialists and developers to pay greater attention to context. The move to abstraction, the new laws seemed to announce, had gone too far. Community and context were regaining importance. During the late twentieth century, this trend would continue, with waves of open-space laws, rules protecting ecologically sensitive lands, and regulations governing historic structures. All of these would take context into account, transforming the rights of ownership in the process. As in the case of the earlier swing toward abstraction and industrialization, this backward shift was hotly contested. Particularly harsh resistance came late in the century with the rise of libertarian thought, which wanted to turn the legal clock back to the late nineteenth century. But the backward swing was nonetheless clear.

This story about property—from the tightly ordered, organic days of feudalism, to the atomistic individualism of the late nineteenth century, and then to the mild reaction of the twentieth century—is not a story only about private property. Indeed, the story has as much or more to do with social relations and interpersonal ethics. Feudalism was foremost a social order. It affected people and communal life as well as land. All people, but particularly women, children, servants, and slaves, had places, roles, and restrictions on what they could do. When historian Louis Hartz in the 1950s penned his influential interpretation of American history, he claimed that America from its beginning was fundamentally a liberal culture.[7] Hartz used "liberal" in the classical political sense of an ideology that honored the individual and sought to free her from undue restraints. The American Revolution, Jacksonian democracy, the antislavery rhetoric of the Civil War, the post–Civil War amendments to the Constitution: all were important milestones in America's saga of individual liberation. In law, this liberal trend was evident in the Married Women's Property Acts of the antebellum era, the end of property-owning requirements for voting, the end of indentured servitude, and a variety of New Deal–era and later laws constructing an individual safety net. In law as in society, individuals were counting for more, and the differences among them—male or female, married or single, landed or landless—were losing importance. The legal individual was becoming an abstraction, stripped of his or her identifying traits. This abstraction often appeared in the form of rhetoric about individual rights, which by the latter half of the twentieth century had spread throughout American culture. A wide range of social problems were framed in terms of how they affected the individual as such, especially whether they diminished individual rights. Abortion was an issue of individual rights, as were health care, welfare, access to public accommodations, and many other social issues.

Hands down, the United States is the home of individual rights and of rights rhetoric.[8] Of course, there have always been dissenters: people who resist reducing matters to the isolated individual; people who stress the importance of social context, the family, the neighborhood, the community, and the collective good. Feminist scholars have been particularly vocal in insisting that relationships and contexts count. The individual, they point out, is dependent on other people and on communal structures for his or her well-being. In the political sphere there is the movement known as communitarianism and various calls to promote civil society. On social issues, fundamentalist Christians have dissented strongly from the atomism of modern culture. Serious conservation thought—promoted by environmental philosophers and important public voices—has also emphasized the communal context. It stresses the vitality of ecological interconnections and the moral status of nonhuman life and future generations.

This history of Western social structures, from the organic, tightly knit, patriarchal, racist social order of centuries ago to the atomism of modern culture, is critical to the ways we talk about moral right and wrongs. The rhetoric of individual rights now holds sway. It has brought huge benefits to Americans as a people, and it needs to bring more benefits to peoples around the world. Indeed, the idea of individual human rights is an extraordinary human invention.

And yet, having acknowledged this importance, we need to admit also that the ideal of the liberated individual is not without problems. Self-constructed visions of the good life (the liberal ideal) can prove troubling to neighbors and community members. People can be selfish and brutish when left free to act as they please. What one person does routinely affects other people, particularly landowners. As populations increase, economies becomes interdependent, cities grow, and a person's well-being depends more and more on the good behavior of other people. Our happiness does not come simply by being set free. Our lives are in fact interdependent; relationships count. Rights rhetoric is a powerful tool to strike down hierarchies, racial orders, and other unfair limits on individual growth. Taken alone, though, it does not set a clear path to the good life.

The field of ethics today is vast and complex, making generalizations difficult. It is nonetheless possible to tease out a long-term course in Anglo-American thought, along a trajectory roughly similar to the one property law has followed. At one time, the individual was knitted into a structured social order with behavior limited by role. The pendulum then swung far in the direction of individual liberty, as generations of reformers attacked legal and social restraints. Now we find the pendulum modestly swinging back as people sense that individualism has gone too far. In small but significant ways, moral thought is paying renewed attention to context and community.

This leaves the subject of ecology, considered both as a science and as a cluster of cultural values about humans and nature. Scientific ecology is far younger than ethics and private property. Necessarily its path has been shorter. Still, one sees in its history a course not unlike that of property and ethics. This path is easier to trace when we define ecology broadly, so that it reaches beyond science to include prescientific and popular ideas about the same basic questions. Here again, we look for long-term trends.

Ecology's story is best told by historian Donald Worster in *Nature's Economy*.[9] This detailed survey explores the various ways people have understood the natural order, how it functions, and how we fit within it. According to Worster, ecology as thus defined has been dominated in recent centuries by a tension between two opposing views of nature. In one view, nature appears in holistic terms as an integrated system, finely crafted and infused with meaning. It is a natural system that we might explore, learn from, and alter, but ultimately one that places limits on our behavior. Now recessive, this perspective on nature long ago was ascendant to the point of being exclusive. Stone Age peoples apparently embraced it fully. In the medieval West, it showed up as the Great Chain of Being. Parson Gilbert White, eighteenth-century author of *A Natural History of Selbourne*, followed in this tradition, as did Thoreau and his fellow Romantics.

Just as the organic unity of feudalism gave way to the new order of individualism, so too did holistic ecology give way to an alternative view of nature, one that was more individualistic, manipulative, and domineering. Encouraged by market forces, Western culture increasingly saw nature as a collection of parts rather than as an integrated whole. Value lay in the parts, rarely in the functioning natural whole. Some parts were valuable, but most were not. Market thinking accentuated this shift. Land was valued chiefly for what it would bring at sale rather than for its use value as a family homestead.

By the time conservation arose late in the nineteenth century, this nature-as-fragmented-resources perspective was dominant, Romantic writers notwithstanding. Nature was "red in tooth and claw," with Darwinian evolution leading over time to increased fitness. Nature was about fierce competition among organisms, not about cooperation and interdependence—at least in the popular mind. Progressive Era conservation ideas tended to soften this atomistic, competitive view. Nature's trees, grasses, and waterways were interconnected, Gifford Pinchot and others explained. The wise way to manage nature was to take account of these interconnections. Still, it was the individual resource that remained central for Pinchot. Nature was valuable because of its parts; thus nature's processes were important only to the extent they sustained the flows of the parts.

A more forceful challenge to this atomistic, market-based view of nature was posed by a contemporary of Pinchot, the plant ecologist Frederic Clements. Clements was interested in nature as a functioning whole and used holistic, even organismic terms when speaking of it. Nature was not a collection of distinct pieces, he explained; it was a functioning unit, with parts that resembled the organs within a body. Popular writers John Burroughs and John Muir also presented nature holistically. The nature they saw, studying it closely, was less individualistic than the popular version. Species were dependent upon one another, unconsciously interacting so as to allow virtually all species to endure. This more organic vision gained in scientific complexity when ecologists Arthur Tansley, Aldo Leopold, and others used systemic terms to talk about landscapes with people in them. Leopold, in particular, put people into this natural whole. He portrayed them as citizens of the overall land community, not conquerors who stood apart. As he did so, Leopold nudged ecology back toward where it had been a century or more earlier, when humans fit into an integrated natural order. Nature was not as tightly composed as an individual organism, Leopold explained, but its ecological processes were nonetheless vital. All life depended upon them.[10]

Clements's holistic vision was influential among ecologists but had little popular effect. In real-world management practices, the atomistic view of nature remained in charge. Leopold's more sensitive understanding of land in the 1930s and 1940s also had only modest public effect. Only after World War II did advocates of ecological individualism meet real resistance. Only then did popular ideas of nature as mother, and nature as carefully crafted whole, stimulate environmental laws that challenged what big business could do. In a crude way, American thought during the era was reviving interest in an older, organic understanding of the human predicament. By late century, the average citizen was coming to see wetlands and other ecologically important lands in new, more ecological ways. These lands were vital nodes in natural systems—parts of integrated natural systems—and best used by leaving them alone. That view of wetlands was a far cry from the prevailing view a century earlier, when draining swamps was a moral crusade.

In retrospect, it is unsurprising that the environmental laws enacted between 1965 and 1980 stimulated a reaction among defenders of the opposing, atomistic view. This happened not just in lawmaking arenas, where industrial interests resisted new regulations, but even within ecology as a science. A wave of ecologists began claiming that nature was, after all, mostly a collection of competing parts, not an integrated whole. It was always in flux, they said, always subject to disturbance, and hence (they implied) a proper subject of intensive management. Older, organic depictions of nature, these ecologists asserted, were true only as loose metaphors. Indeed, the metaphors were so inapt, some said, that they should be excluded from professional discussions. Predictably, this newer, atomistic ecology met a ready reception among industrialists and their defenders. Agribusiness particularly loved it, for it countered the idea that nature's systems deserved respect. Like the libertarian political rhetoric arising at about the same time, this new ecology of flux and disequilibrium quickly showed up in pro-market rhetoric.[11]

The story of ecology, then, fits together roughly as follows. Long ago, nature was understood as an integrated, stable whole. People altered nature by planting crops and burning, but they mostly shaped their lives around nature's ways. In the Western world, the forces of modernity bred a more individualistic, competitive view of nature, which saw nature as a collection of parts, valued the parts independently, and gave a green light to human manipulation. That view brought great gains—just as private property and individual liberty brought gains. But the gains here, as in the other two spheres, came with costs. The land itself suffered, as did ways of life rooted in the land. Late in the nineteenth century, a wide-ranging conservation movement rose up to counter this fragmented view. It insisted that we show greater respect for the interconnected whole when we manipulated land. This conservation impulse became more organic and ecological as the century progressed, at least in some of its forms. Among scientists, greater attention was paid to flows of energy and other ecosystem processes that held the parts together and enabled them to function. Thus, in ecology as in the other fields, the pendulum began to swing away from fragmentation and individualism, though again not without resistance.

Having looked separately at these three fields of knowledge and culture, we can now try to draw them together. In all the areas, an attentiveness to the parts helped bring an end to old, holistic understandings and arrangements. The private landowner escaped the feudal order and gained the right to use her land in ways that brought economic gains. Private property became an abstract concept in ways that fueled economic growth. In moral thought, the individual human stepped out of the tight-knit social hierarchy and gained vast freedoms. The individual increasingly became an abstraction, stripped of personal traits and disconnected from nature and neighbors. In ecology, nature's mysteries and its aura of wholeness gave way under science and economic enterprise to detailed understandings of its many parts and to tools of manipulation. This shift facilitated intensive land-use activities, which greatly increased yields of food, fiber, and minerals. In all three fields, in short, these forces of individualism brought much good. And they continue to bring good in many parts of the world, particularly places where private property is insecure and civil rights are shortchanged.

Nonetheless, we in the United States find ourselves—many of us do, at least—realizing that this process has gone too far. The land has suffered, and so has human life. The pendulum has moved a bit back toward community and context, though it needs to swing back more. Families and neighborhoods as such need to gain greater strength. Individuals too often act in ways that ignore the ripple effects of what they do. Fragmented decisions by landowners too often produce landscapes that no one really likes. Atomism remains too strong.

The United States today needs to move back toward greater awareness of community and interdependence. Socially, economically, and ecologically, we need to engage in a responsible reassembly of the parts so that we enjoy the benefits that arise when communities are healthy—all types of communities, social, cultural, and natural. Natural communities are more than the sum of their parts. The parts are interconnected and interdependent, and the land's fertility depends on the successful functioning of ecological processes. So, too, human life depends in important ways on communal structures, both in meeting basic needs and in producing landscapes that are pleasing and life sustaining. Our challenge, then, is to find effective, responsible ways to rebuild these communities, within our minds, within our cultural values, within society, and on the land itself. We need a responsible communitarianism that respects the values of the parts as such but recognizes also that the parts work best when situated within sound communal orders.

When talking about reassembly and the need for it, it is important to understand what it means and why it requires careful work. The atomism of modernity has long brought calls for unity, sometimes destructive ones. Hitler's National Socialism was a call to reassemble the pieces of German culture smashed by the new industrial order and World War I. Stalin's Communism involved a similar attempt. In the contemporary world, the assertion of radical Islam is yet another such move. We need not look to distant cultures, however, to find such warnings. The cultures of many big businesses and big-city professional firms can assemble people in ways that crush the spirit with their long hours, dress codes, and strict behavior standards. Calls for patriotism and national unity can impose their own forms of flag-draped oppression. In the ecological realm—to shift ground—one might interpret the yearning to reestablish wilderness as a reaction against ecological individualism and a yearning for a distant age of natural holism. As we engage in reassembly, in short, we must never forget the many gains that individualism has brought, even as we admit that too many communities are now sick.

For a small but revealing example of responsible community building, we can look to the work that Wes Jackson and his colleagues are attempting at the Land Institute in Salina, Kansas. Jackson runs an unusual agricultural research operation.[12] Its aim is to develop a form of agriculture that mimics the natural whole of the tallgrass prairie while improving it to produce more food. As a vegetative community, the prairie features perennial polycultures, whose benefits include fertility, water retention, pest control, and the efficient use of energy. These benefits were lost when American farmers embraced radical atomism and turned to annual monocultures, fossil-fuel fertilizers, and chemical pesticides. Jackson and colleagues are not out to reassemble actual tallgrass prairies. Instead, they want to build new, high-yielding ecological communities that include the benefits of integrated prairie ecosystems. It is a provocative idea, applicable in settings far beyond agriculture and land use.

We can situate Jackson's work intellectually by comparing it with the two ends of the land-use spectrum. On one is the unaltered prairie, a highly integrated community that is ecologically healthy but does little to meet human needs. On the other is atomistic industrial agriculture, which is highly productive in annual yields but ecologically destructive and at constant risk of failure. Jackson's work fits in the middle. It represents a swinging back of the slow-moving agricultural pendulum, away from the atomism and individualism represented by the industrial approach and partially back to the holism of the native prairie. It is a carefully charted swing back, though, with constant attention paid to human needs, to the limits on knowledge, and to the built-in wisdom of nature itself.

In the case of private property, we could benefit from a similar continued shift, reintegrating the individual land parcel into the social and ecological communities of which it, and its human owner, are parts.[13] Carefully done, such a shift could bring the many benefits of healthy communities without losing the gains that come from secure private rights. One of the central reassembly tasks is to redefine the term "land-use harm" so that it includes harms to the surrounding social and ecological communities as such. When judging the legality of land uses, context needs to count more than it does. Ownership rights in a wetland, in a sloping field, and in critical wildlife habitat ought to differ considerably from ownership rights in a flat, dry Illinois farm field. Put otherwise, nature and context deserve greater roles in defining the particular rights that landowners possess. On the good side here, we already have the ideas we need to undertake this kind of responsible reassembly; we know in basic terms how property norms should change. Unfortunately, the steps required to implement these ideas are at the moment politically and culturally infeasible. And they will remain infeasible until we take time to probe private property as an institution, recognizing its flexibility, reflecting on its philosophic implications, and accepting our moral responsibility for how it functions. Further movement toward community will require hard work.

In the case of ethics, our situation is strikingly similar. We need to recognize that maximum individual liberty simply makes no sense as a guiding ideal. This is particularly true when we define liberty in negative terms, as freedom from restraint, while ignoring the kinds of positive liberty that are required to promote healthy landscapes, harmonious neighborhoods, and other collective goals. Individual freedom is good, all things equal. But all things are not equal. Healthy landscapes; curbs on urban sprawl; restored waterways; vigorous, diverse wildlife populations—all require collective action of a type that clashes with libertarian ideals. To achieve these good landscape goals, people have to coordinate their individual actions so as to promote the common good. Working together, they need to restrain themselves.

These various kinds of community building will not come easily. The basic ideas we already have, but the challenge arises because the forces of individualism remain so powerful, culturally, economically, and politically. And they will not give ground easily. In ecology we have the forces of industrial farming, forestry, and ranching, which control vast acres and research dollars. Aiding them in resisting change are proponents of the nature-as-flux school of thought. Our ecological metaphors are causally linked to cultural values outside science, in ways rather easy to see. It is thus no coincidence that the rise of libertarianism in politics has been matched by a similar rise in the nature-as-flux view of nature.

In the area of private property, community advocates today collide with legions of free-market enthusiasts and intensive land users. Within law schools, Langdell's chemistry-based form of legal abstraction has enjoyed a bit of a resurgence. Its many modern proponents show enthusiasm for various reductionist approaches to law, particularly the perspectives known as law-and-economics and contractarianism. In the ethical realm, the call remains strong for even greater individual rights. The call these days is perhaps loudest on the far right of the political spectrum, and it is being pushed by powerful economic forces that make money in a market without limits. Madison Avenue advertisers plainly have a dream, and it is of the amoral individual consumer who is willing and able to buy and consume without concern for others. This is hardly the only form of radical individualism. A related one, quite different politically, is the animal-welfare movement, which sees moral value in the individual animal rather than in ecological wholes.

We can sum up. Running through ecology, ethics, and private property is a distinct fault line. On one side are those who continue wielding the rhetorical tools of abstraction and individualism; on the other, those who see clear needs to recognize context and to rebuild communities. For the latter view to gain ground—as it needs to for lands to become healthy—its proponents need to do their intellectual homework. They need to get clear on what they are doing and how best to talk about it. They need to think imaginatively and construct alluring visions of what life can be like. They need to excite the hopes of citizens.

People who care about context and community have a cultural battle on their hands. It is the battle for community: for pleasing neighborhoods, healthy rivers, sound local schools, and ecological integrity. The opposing forces of individualism succeed by division. Better than they have, the forces of community and reassembly must work by cooperation.

Notes

1. My discussion in this paragraph and those that follow, about Langdell, Eliot, and law as science, draws chiefly upon the following works: Albert J. Harno, *Legal Education in the United States* (San Francisco: Bancroft-Whitney, 1953); Robert Stevens, *Law School: Legal Education in America from the 1850s to the 1980s* (Chapel Hill: University

of North Carolina Press, 1983); Dennis R. Nolan, ed., *Readings in the History of the Legal Profession* (Charlottesville, VA: Michie, 1980); Arthur E. Sutherland, *The Law at Harvard* (Cambridge, MA: Belknap Press, 1967); William LaPiana, *Logic and Experience: The Origin of Modern American Legal Education* (New York: Oxford University Press, 1994); Anthony Chase, "The Birth of the Modern Law School," *American Journal of Legal History* 23 (1979): 336; Marcia Speziale, "Langdell's Concept of Law as Science: The Beginnings of Anti-Formalism in American Legal Theory," *Vermont Law Review* 5 (1980): 27; Howard Schweber, "The 'Science' of Legal Science: The Model of the Natural Sciences in Nineteenth-Century American Legal Education," *Law and History Review* 17 (1999): 421; Michael H. Hoeflich, "Law and Geometry: Legal Science from Leibniz to Langdell," *American Journal of Legal History* 30 (1986): 95.

2. Quoted in Sutherland, *Law at Harvard*, 175. Howard Schweber contends that the standard story about Harvard's embrace of legal science and the case method emphasizes Langdell too much and Eliot too little; as Schweber notes, Eliot was, at least in public settings, the more frequent defender of the case method of study and analysis. Schweber, "'Science' of Legal Science," 458.

3. Quoted in Stevens, *Law School*, 38.

4. Quoted in ibid., 40.

5. This story is told in more detail in Eric T. Freyfogle, *The Land We Share: Private Property and the Common Good* (Washington, DC: Island Press, 2003), 52–90.

6. Eric T. Freyfogle, "Community and the Market in Modern American Property Law," in *Land, Property, and the Environment*, ed. John F. Richards (Oakland, CA: ICS Press, 2002), 382–414.

7. Louis Hartz, *The Liberal Tradition in America* (New York: Harcourt Brace, 1955).

8. Critiques of this cultural tendency are offered in Mary Ann Glendon, *Rights Talk: The Impoverishment of Political Discourse* (New York: Free Press, 1991), and Robert N. Bellah, Richard Madsen, William M. Sullivan, Ann Swidler, and Steven M. Tipton, *Habits of the Heart: Individualism and Commitment in American Life*, updated ed. (Berkeley: University of California Press, 1996).

9. Donald Worster, *Nature's Economy: A History of Ecological Ideas*, 2nd ed. (Cambridge: Cambridge University Press, 1994).

10. Leopold's scientific ideas are explored and situated in the flow of ecological thought in the 1930s and 1940s in Julianne Lutz Newton, *Aldo Leopold's Odyssey* (Washington, DC: Island Press, 2006).

11. The apparent cultural origins of this new ecology are considered in Donald Worster, “The Ecology of Order and Chaos,” in *The Wealth of Nature: Environmental History and the Ecological Imagination* (New York: Oxford University Press, 1993), 156–70.

12. See Wes Jackson, *New Roots for Agriculture*, new ed. (Lincoln: University of Nebraska Press, 1985), and Judith D. Soule and Jon K. Piper, *Farming in Nature’s Image: An Ecological Approach to Agriculture* (Washington, DC: Island Press, 1992).

13. I consider various options in Freyfogle, *Land We Share*, 203–81.

7

LOVE AND DEMOCRACY

One of Wendell Berry's valuable contributions to conservation thought has been his persistent reminder that good land use rarely occurs in isolation. Good land use requires human users who care about land and know how to use it well. Sound knowledge is critical, and so is a supporting social order, a community that can inform and help sustain good practices. Such a community, in turn, requires leaders if it is to respond adequately to tensions and challenges. Not all local people but enough of them must display a commitment to the community and a willingness to serve. In Berry's view culture counts, and good culture requires defenders.

Berry's major novel from 2000, *Jayber Crow*, exposes some of the challenges involved in living ethically within a small community. The book's extraordinary protagonist is the barber of a tiny town along the Kentucky River. Guided by New Testament gospels, he strives to comply with the tenets of Christianity, especially the admonition to love all his neighbors. For barber Jayber Crow and, presumably, Berry, Christian ethics have to do chiefly with relationships, among people and between people and nature. To love people who are virtuous and who return that love is relatively easy. Far harder is to love the unlovely and unloving, particularly pride-filled people who degrade the local order. What does a person do when confronted by neighbors who respond to the sirens of greed? How should we deal with people who love themselves and not the created land?

In *Jayber Crow*, Berry ponders these issues as they are confronted by the protagonist and his colorful friends. His novel expressly invites comparison with another morally laced river book, *The Adventures of Huckleberry Finn*. Just as Twain did, Berry precedes his novel with a warning to readers not to search for meaning within the text. Immediately we are alert to the similarities and differences between the two works. Both books have to do with liberation, companionship, and the sustaining and corroding forces of culture. In both, the river provides a route of travel. Huck Finn and Jim follow their river to escape small-town life; Jayber Crow uses the Kentucky River instead to end his youthful wanderings and find his way home. In a sense, Berry has written a sequel to *Huckleberry Finn*, proposing a morally superior conclusion. Twain's book stumbles toward the end (as commentators have long noted) because of uncertainties about what Huck should do next. Having learned what he set out to learn, the wandering Huck might have returned home to join and help improve his riverside hometown. But Twain had difficulty imagining a settled, small community that was not confining and corroding, and so Huck is dispatched to the territories. What if, instead, Huck had chosen to return home? And what if, upon returning home, he had taken seriously the Christian gospels and tried to love his fellow town members, all of them?

Unlike Huck, Jayber Crow does return home, or very close to it—to the richly imagined fictional world of Port William, Kentucky, which Berry has crafted in book after book. Port William, as Berry realizes, is on a slow downward course, as are countless similar towns. Jayber recognizes this too, yet he nonetheless casts his lot with the town and its people. He does not expect to reverse the town's fortunes. He wants only to become a responsible member who sustains the community as long and as well as he can. To do that, Jayber works hard to sustain good relations with the people around him. What we have, then, is a character highly unusual in modern fiction. Jayber intentionally enters into a communal order rather than escaping from it. And he sees his plight and calling not in isolation but intermingled with the lives of his neighbors. Watching Jayber as he ages, we consider what community membership is all about.

Given Jayber Crow's serious moral quest, it is appropriate for us to assess his achievement. The story opens in 1986 when Jayber is seventy-two years old. We learn of his life mostly through flashbacks. Did he become a responsible community member? The answer, in many respects, appears easy. Despite a few slips, he has lived a life of uncommon virtue. We are amazed by his humility, kindness, and willingness to serve. Yet there is something distinctly lacking in Jayber, in his town, and in the agrarian moral order that he upholds.

Jayber's individualism is highly moral, but he is an individual all the same. And he views other people as individuals as well. In his mind and heart, he is a community member, but that membership rarely translates into concerted action with other people. We watch and admire Jayber as family member, church attendant, neighbor, and friend. What we do not see, not ever, is Jayber acting as citizen. He abides by the law yet keeps his distance from government. We know his political type, of course. He is a Jacksonian democrat, happy when government is small and unobtrusive. In Jayber's world, individuals ultimately are free to make their own choices. Constraints, if any, come from the moral order, not from law. As we watch Jayber, we are prone to wonder, How can a community deal with misfits and troublemakers when good people like Jayber stand aside and do little? How can it care for its common wealth if community leaders stand strong but silent?

This problem is particularly severe in Berry's novel because Jayber's respect for individual freedom goes even further. Even in the face of behavior that he despises—bad land use that saps the town and its future—Jayber remains mute, unwilling to criticize openly. Jayber does not protest the destructive ways of the world by organizing for change or even by speaking out. He opposes them by embracing a personal life based on opposing values. Meanwhile, though, the agents of destruction proceed apace. Indeed, the town's chief bad actor remains unaware, year after year, that Jayber despises his values and achievements.

With his unusual tale, Berry poses a critical question. What does it mean to be a good community member, given the links between community and land use? Is Jayber right to rest his hope in the slow spread of individual virtue, or as readers can we protest his failure to speak out and resist? Is love the true key to healthy communities, as Jayber thinks and Berry proposes, or do we need a dose of strong democracy as well?

Jayber Crow entered life in August 1914, one day before the beginning of World War I. This timing signifies what soon becomes apparent to us, that Jayber is linked to an older age and its values. If he is not the last of his kind, he is on the tail end. His real name is Jonah; his birthplace, Goforth. We anticipate, then, his role as reluctant prophet, dispatched by storm to some Kentucky Nineveh carrying a call for repentance and warning of judgment. Jayber's parents die when he is young, and he is sent to live with an aunt and uncle two miles distant. They also die, by the time he is ten, and he goes next to a Christian orphanage. Like Huck, Jayber feels confined. His yearning, though, is "not to go into the town at night but to escape into the countryside in the daytime." He is drawn to pastoral landscapes, to rolling farmland and wood-lined streams. As he listens to his dogmatic orphanage teachers, Jayber is troubled by their elevation of the soul over the body and by their claim that order comes from human institutions, disorder from nature. His own interpretation is quite different.

Jayber soon feels called to the ministry and moves to a small denominational college in the town of Pigeonville. Freed from confinement he is tempted by the allures of the world. "Call me J.," he tells his new acquaintances, echoing the famous character who used the same line, Jay Gatsby. For the moment at least, he wonders which prophet to follow. Jayber, though, takes the gospels seriously—too seriously, perhaps, when he suggests that his minister-instructors are failing to live up to the teachings of Jesus. How can they support war when Jesus demands love, he asks? And if God so loved the world, then should not Christians love it also, all of it? Jayber struggles, too, with prayer, and wonders how a Christian could expect or even ask God to take instructions. Surely the only rightful prayer, he concludes, involves passive submission to God's inscrutable will.

Early in 1935, twenty-year-old Jayber leaves the church college and heads to Lexington. The Depression is deep, and he has trouble making his way. He locates temporary work at a horse-racing track, symbol of the worst traits of the modern era. At the track, horses and riders race in circles and money goes to the swiftest; speed is all and direction unimportant. Jayber soon takes hold at a nearby barbershop and slowly learns the trade. Meanwhile he is drawn to books and attends classes at the University of Kentucky. But his journeying is not done; he has not yet found his place.

In January 1937, now twenty-two, Jayber decides to leave his post and studies and to head toward the bigger city, Louisville. A major flood is unfolding, and he wants to see it. It is a dark Friday. All day he journeys on foot in the ceaseless heavy rain. By evening he reaches Frankfort and the greatly swollen Kentucky River. The harsh weather brings to mind his deceased aunt. "It was exactly what Aunt Cordie could make you imagine when she was in one of her end-of-the-world moods—the signs being fulfilled, and the dreadful horsemen about to make their way across the earth." Jayber wants to reach the other side—a promised land?—but the bridges are all closed. He talks his way around a police officer standing guard and then eases his way up the aptly named St. Clair Street bridge. From there Jayber surveys the formless waters. It is not the end-time, he senses, but instead a time of re-formation. Over "the troubled surface of the water" he sees "a strengthless, shapeless cloud of light." Passages from Genesis come to his mind. The earth was without form, void and dark, with creation unfolding. He reaches the far side as the waters rise, and he is soon directed toward the state capitol, looming high. It is the ark, and he and other survivors will stay there until the waters recede. Jayber realizes he is now on the right path, heading not to Louisville but home, to Port William. His personal resurrection begun, Jayber wanders the next day—Monday—onto terrain that he recognizes from his youth. "It just all of a sudden came over me," he recalls years later, "that in one breath I was lost and a stranger, and in the next I was found."

Jayber quickly settles in to Port William. He takes over the abandoned barbershop with a studio residence above and becomes part of town life. By choice and disposition, he adopts a decidedly Christian role at the bottom rung of the social order. He is not a washer of feet—the paradigm of Christian service—but he is as close as he can be, living monk-like and meeting an elemental human need. His barbershop door is open to all, without regard for station or vice. It is a life, we recognize, as integrated and humble as it can be. Behind his small building is a patch of earth, which he gardens out of passion and necessity. The earth, too, is part of his monastic enclave.

Within a few years, the new war begins—or, as Jayber and his neighbors view it, the long-simmering war revives. Jayber must decide what to do. He realizes then and there that he has made his choice. He cannot imagine a life for himself beyond Port William. He will share his fate with the fate of the town. A heart murmur (a sign of Jayber's sensitive spirit) keeps him from the fighting. He will remain home instead, amid the families who fear and experience the loss of sons. Because his behavior is a bit wayward—on occasion he attends drinking parties or hangs out with stray women in a larger town—Jayber the bachelor is viewed with suspicion by Port William's righteous churchgoers. Still, he fits readily within the bounds of the local church and is hired to clean the building and dig graves. For the most part, he remains at arm's length from the church because the young ministers who come and go have little feel for the community and its land. These transient ministers, like his teachers at the orphanage and church college, dispense dogma without recognizing the gospels' high demands. They rarely respect God's creation.

When the modern age enters Port William in the form of rigid health regulations of barbershops, Jayber decides to distance himself even further from it. The year is 1968, Jayber is fifty-four, and wars, riots, and protests dominate the news. Unwilling to spend money installing hot running water, Jayber closes his shop and follows a course of downward mobility. He takes up residence in a cabin by the river, surrounded by wilder nature. More of his food will now come from fishing and gathering. To his surprise, his friends follow him there. He cannot lawfully cut hair, given the health regulations, so he shifts from the cash-contract economy to an earlier, more solid world of barter and gift exchange. Now, even more, the nearby river looms in his imagination.

Jayber is never tempted by marriage, nor are his prospects strong. What he develops is one of the most extraordinary loves in modern fiction. Jayber spots the young daughter of a community-leading farmer, Athey Keith, and gains a fondness for her from afar. He watches as she matures, becoming an exemplary woman and community member. The girl, Mattie, takes an early interest in a self-absorbed, athletic, handsome youth of similar age, Troy Chatham. She is too young for Jayber and beyond reach. He can only watch her, carefully and often. Berry's novel is about love in its many forms, and Jayber's infatuation with Mattie goes to extraordinary lengths. In time it becomes the ideal chivalrous love of the medieval world—the love of the virtuous courtier who devotes himself to a high lady whom he occasionally sees but cannot hope to attain.

One sleepless night Jayber wrestles with his extraordinary love and decides, knight-like, to seal it with a vow. Alone in the dark, no one to hear, he pledges his love for her. He vows to be the faithful husband of a woman already married. Jayber will never speak of his pledge, nor can he hope to be her husband in the flesh. His love is pure and selfless; he has reached an exceptionally high plateau. Years later, he will occasionally encounter Mattie in a forest patch, and they will exchange brief words. Yet even on her deathbed, when Jayber overcomes his reluctance and visits her, he can only refer to his feelings obliquely.

Jayber's disagreement with the modern age centers on Mattie's husband, Troy—aptly named because of the disruptive values and industrial tools that he brings into peaceful Port William. (He reminds us, too, of a dashing, shallow character in Hardy's *Far From the Madding Crowd.*) Troy is the emblem of a more selfish, money-driven culture, willing to gamble to get ahead and confident in his dismissal of old ways. Troy marries Mattie—the dynamo joined with the virgin. Immediately he presses his father-in-law for more land and greater freedom in using it. Where Athey Keith uses caution, addressing his farmland with respect, Troy is anxious to press it hard to produce. For Athey, "the law of the farm was in the balance between crops (including hay and pasture) and livestock. The farm would have no more livestock than it could carry without strain. No more land would be plowed for grain crops than could be fertilized with manure from the animals. No more grain would be grown than the animals could eat." Troy, in contrast, "thought the farm existed to serve and enlarge him." He was anxious to bring on the tractors, expand the acres, buy more inputs, ignore the balance, and press the land to yield cash. His was a new way of farming, dependent "not on land and creatures and neighbors but on machines and fuel and chemicals of all sort, *bought* things, and on the sellers of bought things—which made it finally a dependence on credit." Only slowly can Jayber and others see that "this process would build up and go ever faster, until finally it would ravel out the entire old fabric of family work and exchanges of work among neighbors."

Year by year, Jayber watches as Troy takes over more land, works harder, presses the land, goes into debt, buys more equipment, and distances himself further from the values of his father-in-law and silent wife. Indeed, Troy can hardly imagine the life of his in-laws, seized as he is by "a daydream of 'farming big'" and overflowing with "the impatience of the new." Despite his bravado, though, Troy is a lonely man. "He was lonely because he could imagine himself as anything but himself and as anywhere but where he was. His competitiveness and self-centeredness cut him off from any thought of shared life." While Mattie plays the dutiful Christian wife, Troy is intent not on magnifying the Lord but on "magnifying himself with power." The result is suffering, imposed on the land and on Troy, Mattie, and their children.

While fulfilling her wifely role, Mattie receives from her parents legal title to a tract of forestland that has long gone uncut. Its value as timber has risen, and Troy is anxious to harvest it to gain cash to pay debts and expand. Mattie views it as her parents viewed it—as a safety net, her nest egg, to use only in the hardest of times. Quietly she resists Troy's pleas to sell the timber. When Mattie is on her deathbed, Troy promptly sends in the saws and tractors.

Jayber Crow views Troy in the same way that Athey Keith does, as a repudiation of much that is vital. Troy goes it alone, showing little interest in the community. He presses the land hard, confident that he can make it do what he wants. His practices and guiding values undercut Port William's vitality and future. The town is sliding down, its stores gradually closing, its young people disappearing, its farm incomes weakening because of market forces beyond its control. Port William, Jayber could see, was "like a man on an icy slope, working hard to stay in place and yet slowly sliding down." Troy, no doubt, is not the only farmer giving in to the market's lures. But he is the one we watch, and so the treatment he receives is revealing. As Troy boasts and schemes, the good community farmers remain silent. "He was looked at and listened to without comment," Jayber explains. Troy comes to Jayber for haircuts and leaves with little awareness that Jayber opposes his world. Troy launches monologues about his operation and finances, Jayber relates, "without suspecting in the least that I did not concur in his high opinion of himself. . . . I acted toward him with tolerance and politeness."

Port William hardly resembles Nineveh in its iniquities, yet it does have weaknesses. Jonah "Jayber" Crow could have spoken out against them, rather than submit silently. Like Huck Finn, Jayber rejects those parts of culture that he dislikes. He opposes them by refraining from participating in them. This character trait appears often in Berry's fiction, and in a favorable light. It is associated with the characters that Berry presents most warmly—the town leaders such as Mat Feltner and, here, Athey Keith and Jayber Crow. When a barbershop patron utters an overtly racist remark, Athey Keith challenges him. But the rare challenge establishes merely the outer bounds of acceptable public behavior. It does not begin to identify and inculcate good behavior. Among Berry's frequently recurring characters, only the lawyer, Wheeler Catlett, challenges other people regularly, and he does so mostly in settings where he acts as lawyer and advisor, not merely as town member.

We see this reticence respected also in Berry's succeeding novel, *Hannah Coulter*. Nathan Coulter, one of the good farmers, has returned from World War II with senses of loss and absence, wanting to create a small place where he can live apart from the urges and forces that brought the war about. Years later, looking back to the postwar period, his widow Hannah explains, "There can be places in this world, and in human hearts too, that are opposite of war. There is a kind of life that is opposite to war, so far as the world allows it to be. After he came home, I think Nathan tried to make such a place, and in his unspeaking way to live such a life."

Hannah Coulter is talking about her longtime husband. We should take note then when Hannah says "I think." Her qualification makes clear how "unspeaking" Nathan was in his pacifism. Jayber, too, sought to create a "place in this world" that was opposed to the ill trends of the modern age. His final home along the river was one such place, but he succeeded even more in creating a private place within his heart. He too was unspeaking.

In another work, Berry fondly describes a real-life riverside place that was opposite to war in his biography of painter-shanty-boater Harlan Hubbard. Harlan and Anna Hubbard retreated from the warring world to create an enclave of health and inner wealth along the Ohio River, not far from the mouth of the Kentucky. Berry knew and admired the Hubbards. In his book about them, he recalls how, years earlier, he had lamented Harlan Hubbard's failure to attend a rally to protest a nuclear power plant, under construction within sight of the Hubbards' home. It later occurred to Berry that Harlan did protest the power plant—by living a life opposite to it, a life disconnected from the power grid and from much of what it represented.

In the case of barber Jayber Crow, it is appropriate to evaluate his behavior given his deliberate decision to mix his life with the fate of his town. What are we to make of Jayber's silence in the face of degradation? The question, of course, is not about literary value. Jayber is true to his times and in the novel cannot be other than he was. The question is whether, transported from fiction to history, different behavior might have produced better results. Should he have behaved differently, and should his successors on the land and in small towns today behave differently?

Jayber's concerns, viewed from a distance, have to do with

proper respect for relationships. In his personal affairs, Jayber attends to them constantly, doing his best to forge bonds and passageways in the appropriate ways. He is a good neighbor, good friend, fair businessman, and respectful church member. He pays homage to those who have gone before—to those he carefully buries and the inhabitants of the graves he tends. He remains circumspect within his barbershop to ensure that social interactions there go properly. And then there is his most demanding relationship—his bond of love with Mattie Chatham, a married woman. This relationship, he knows, can be right only if he keeps his distance, neither seeking her out nor speaking of love. His achievement is extraordinary, indeed otherworldly.

When relations among the parts are right, Berry suggests, the parts form pathways for the transmission of caring, work, love, memories, wisdom, and fertility. The best relations and exchanges are those that entail no accounting. Sharing and free offering rank above the cash sale. Jayber, accordingly, rises to a higher moral level when his barbering shifts from a cash basis to a casual, exchange basis (his customers often leave cash but quietly and without notice). In his attention to relations and bonds, Jayber stands among the Port William residents who enjoy Berry's favor. Old Jack Beecham, we learn in the book about him, succeeded well in all of his marriages in life—to place, neighbor, and tradition—with the overt exception of his marriage to his wife. In Berry's heartfelt tale "The Boundary," we learn that Mat Feltner has managed his farm so as to integrate it successfully into nature, into the surrounding social and economic orders, and into the flow of generations—even as he keeps tight his barbed-wire fences. Proper relationships count, for Berry as for Crow.

Jayber's chief complaint about Troy Chatham, then, deals only indirectly with Troy's egotism. It is that Troy performs poorly in forging and sustaining the relationships that ought to characterize his life. He is a poor neighbor and friend because he listens little and understands less. When he rebels against his father-in-law, particularly in tending the land, he breaks the vital bond linking generations. His abusive practices toward the land itself reflect other failed relationships, between farmer and farm and between farmer and other creatures, as does his unfaithfulness in marriage. We have little direct evidence of Troy as Christian, but we can infer from his behavior that his links here are indifferent. Jayber responds to Troy's failings by working even harder to keep sound all of his own relationships. Troy is his biggest challenge—his most difficult test as he strives to love all his neighbors. Jayber never really loves Troy, but he withholds judgment and remains willing to forgive; he is the father ever ready to embrace the prodigal son.

Aside from the river, perhaps the most powerful image in *Jayber Crow* is the Keith family nest egg—the tract of mature, ecologically complex forestland that Athey Keith has kept in reserve, unwilling to harvest the timber except under duress. The nest egg embodies the constellation of values, affections, loyalties, and restraints that Athey embraces and feels duty bound to pass along to the next generation. To protect this forest is to accept a long-term perspective. It is to exercise caution and to place a curb on appetites, to show respect for wild nature and a desire to keep it close. In Athey's view, the farmer is the less important, more short-lived participant in a sound marriage of people and place. All of these ideas and sentiments come together in the nest egg. Jayber knows it, and so does Mattie. Troy does not, or if he does, he simply does not respect the ideas and sentiments. When Troy unleashes the chain saws upon Mattie's death, he proclaims his willingness to break the bonds. The forest as saleable timber he understands; the forest as cultural complex he does not.

Jayber's passive response to Troy's behavior is consistent with his Christian ideals. Jayber is the paradigm protestant, disinclined to take guidance from institutions. He reads the Bible and makes up his own mind, ignoring most of the Old Testament, embracing the gospels, and excising Paul's legalism. He is on a quest for heaven—indeed, his extended autobiography, he tells us, is a story about heaven. Jayber's Christianity is thus highly individual. The church is a collection of individuals, not a covenanted community. Jayber's economic fortunes plainly depend on those of Port William, and his neighbors' pains become his pains. But when it comes to salvation, he apparently plans to go it alone.

Jayber's individualism—which overlaps, in truth, with Troy's—

highlights this limitation on Jayber's sense of the community. Port William faces two types of worries as it considers its future. There are its worries about its ability to sustain the moral order that upholds the common life, worries that center on marriages, childrearing, neighborly ties, hard work, and personal rectitude. In the essential context of land, they include the community's ability to husband its accumulated wisdom on using local land well and to hand it down in orderly fashion, generation to generation.

Along with these interior worries are those stimulated by the external forces pressing upon Port William. Jayber mentions the Economy and the War—capitalizing both to make clear their breadth. Local producers are price takers and inevitably suffer when commodity prices decline. Distant cities, jobs, and glamour lure the town's young people away. Advertising stimulates demand for the market's wares, which means demand for the kinds of lives and jobs that feature the consumption of these wares. In Troy Chatham we see these external forces hard at work, pulling him to borrow, buy, labor, and exploit the land.

Jayber deserves praise as a community member for his efforts to deal with the first category of community worries. He withholds overt criticism of others, but his personal rectitude is influential. Moral standards are best upheld when a critical mass of people support them. Jayber does less well dealing with the second category of worries. In his personal life, he resists the market's wares and sirens as best he can (he belatedly buys an automobile but then has second thoughts). Beyond that, he remains passive. He does nothing, alone or with others, to counter any of the external pressures. Surely Port William, ably led, could take firmer control of its lands and land uses. When the local school is closed and the town slips down another step, we hear only laments and murmurs from town residents. And of farm cooperatives and co-op stores we hear nothing. Jayber, it seems, has responded to some of the community's threats but not to others. Can we criticize him, then, for being only half of a true community member? Has Berry's Huck returned to his hometown only to fill part of his communal role?

In his fiction Wendell Berry returns often to the need for a community to be aware of itself. Young Andy Catlett (who resembles Berry himself) is particularly mindful of the communal stories and of his chosen duty to learn and safeguard them. Andy's father, Wheeler Catlett, helps connect good young farmers with good land to keep intact the handing down of practical farm wisdom. Mat Feltner, a model farmer and town leader, provides a looming moral presence in matters relating to the family, the household, and public decorum. Jayber's special role is to model the selfless Christian servant. Together, they keep Port William intact morally, or as intact as it is likely to be. The characters do not, however, come together in overt ways to set standards and enforce requirements for community membership. They do not defend themselves against the local manifestations of global forces. Instead, they turn the other cheek, and have it slapped as well.

Jayber clarifies this limitation when describing his barbershop, a common place for men and boys to gather. "My shop," Jayber recalls years later, "was a democracy if ever anyplace was. Whoever came I served and let stay as long as they wanted to." The term "democracy" jumps out here given that politics and governance are nearly absent from Berry's fictional world. Jayber's barbershop is a place where individuals express their views, wise or foolish. This free expression of opinions is what democracy entails, in versions of democracy that view people simply as individuals with self-selected preferences, ready for government to collate. But this is a thin version of democracy, so thin, really, that few political theorists would view it as democracy at all, any more than the free market is a democracy when it aggregates buyer preferences. Democracy is governance by the people collectively, the demos. Strong, effective democracy requires efforts to learn and convince. It works best when people deliberate in settings where they can assemble facts and exchange views. What if Jayber's barbershop had become, after hours, a place for community members to talk about their community, sharing visions and proposing action? What if his clients and visitors had spoken about land-use issues and about steps they might take to protect farmlands, wild spots, and ecological processes?

In his embrace of individualism, Berry is a classical liberal. He shows his libertarian streak in his disdain for all forms of servitude and domination. As a liberal, though, Berry must respond to the charge leveled against early liberalism by defenders of older, organic orders. Liberalism would descend into chaos, critics said, because it lacked a governing order to contain personal vice. Early liberals took this charge seriously. And they had an answer, or thought they did, in Adam Smith's invisible hand. When people were unleashed economically, Smith contended, the magic of the market would transform individual vice into public virtue. Striving to get ahead individually, people would labor hard, lead reasonably orderly lives, and add to the nation's wealth. This was the liberal answer. For a few generations, it sufficed.

This answer, of course, is not one that Jayber and Berry are likely to propose, and with good reason, particularly on land-use issues. Economic theory might predict a happy result, but the facts in real landscapes often speak otherwise. The profit motive does not always lead to good land tending. Land economists could see this plainly a century ago, well before the prominent private-lands disaster of the Dust Bowl of the 1930s. In an important paper in 1913, economist Lewis Cecil Gray explained in detail why the invisible hand of the market was often too weak to encourage landowners to use their lands conservatively. Particularly vivid evidence appears in the historical record from the antebellum era, according to historian Steven Stoll. Perhaps the first major chapter in America's conservation saga began in the decades from 1820 to 1850, when key farm reformers and newspaper editors pushed farmers to use their lands better. They promoted greater rotation, the better use of organic manures, and other restraints to keep land from becoming "exhausted." The effort largely failed, and it did so, Stoll concludes, because the sound methods of land use simply made little economic sense. Market forces rewarded landowners who abused their lands and then walked away. For all we know, Troy's decision to cut the nest egg made equally good economic sense. Berry is wise to withhold support for Adam Smith's reasoning. But without it, we are back to the longstanding complaint: liberal individualism can bring chaos and decline unless mechanisms exist to contain disruption.

Berry issues a call for individual rectitude. He proposes that individuals think more about their duties than about their individual rights. But are we left, then, merely to hope some day for a stateless society, a community in which people equate personal interest with social good? Are we to assume that the interests of individual and community really are the same, and that people would see this if only they discarded their false consciousness? Is the dream embedded in Berry's fiction similar to the one of Karl Marx, who predicted that states would become unnecessary once workers recognized their solidarity and brought personal interests into line with the collective good?

What we have in Berry's fiction, by implication, is a version of the liberal state as night watchman. Individuals are free to act as they see fit. They can choose among competing visions of the good so long as they do no violence to others. The state takes no stance among these moral visions except by barring crime. Proponents of this view often describe it as morally neutral, but it is not. To give similar treatment to two types of behavior is to suggest they are morally equal. When the law allows Troy to farm as he does and Athey to farm differently, it labels both approaches socially acceptable. Such a minimal state plays directly into the hands of the economic forces pressing down places like Port William. The Troy Chathams of the world are likely to embrace this economic liberalism, just as they embrace ideas about private property in which owners are free to act as they please. Both give a green light to their selfish urges.

Jayber Crow's silence is consistent with his character and his times. But let us imagine an unliterary revision of Berry's fine work. What if Crow and his like-minded neighbors instead had talked about their shared landscapes, maybe even about the nest egg in particular? Could they have taken action to keep it from the auction block? Could they have tried to buy the land or some interest in it, or pressed for county riparian-corridor zoning or a state forest-practices statute? Ignoring cultural barriers (which Berry the novelist could not do, of course), the good members of Port William could have proposed various ways of protecting the nest egg as forest. Had they done so, they might have also clarified and protected the cultural values that the nest egg represented. Values are strengthened and shared when people labor together.

Standing on the fringe of his Port William church, Jayber Crow could see clearly enough what brought its parishioners together, week after week. They came to acknowledge "their losses and failures and sorrows." They also came to express "their wish (in spite of all words and acts to the contrary) to love one another and to forgive and be forgiven." What was true of the gathered church was true also of the larger Port William community. "Imperfect and irresolute," it was "held together by the frayed and always fraying, incomplete and yet ever-holding bonds of the various sorts of affection." In his imagination Jayber could draw that community together and see it clearly. "The community must always," Jayber could see, "be marred by members who are indifferent to it or against it, who are nonetheless its members and may be nonetheless essential to it. And yet I saw them all as somehow perfected, beyond time, by one another's love, compassion, and forgiveness, as it is said we may be perfected by grace."

Here we see where Jayber ends up, as he watches Troy and others like him mar his small community. Somehow, Jayber hopes, they can become perfect not in the here and now but in some other realm, beyond time, by love and forgiveness. Love is the ultimate bond and pathway. If the relations are perfect, Jayber assumes, all will be well. In the meantime, though, the rare forest is cut, and the cultural bonds it represents are severed. As for Jayber, he remains strong in his love and silence as he watches Mattie die. By staying his course, living humbly and keeping his vow of selfless love, he has fulfilled the poetic prayer of the biblical Jonah:

> Those who worship vain idols
> forsake their true loyalty.
> But I with the voice of thanksgiving
> will sacrifice to you;
> what I have vowed I will pay.
> (Jon. 2:8–9a, New Revised Standard Version)

What Jayber overlooks is that love comes in many forms, some tougher than others. Could he have loved Troy Chatham while also confronting the man and challenging his ways? Could he have loved Troy even as he urged the county government to curtail unwise timber harvesting? Does love necessarily mean standing by silently?

Jayber is plainly an advance on Huck Finn in terms of his commitment to place. Yet can we hope for something better still, for a type of community member who advances beyond Jayber, for someone who speaks forcefully about good land use, ecological interconnection, the social origins of private property, the economics of conservation, and related topics? Can we hope, in the case of Port William, for an activist community visionary, guided by different understandings of democracy and liberty, someone who can help community members take charge of their fates?

8

WANTED: ENVIRONMENTAL LEADER

The United States is currently seeking one or more national environmental leaders. Applications for the position are invited, especially from individuals, resident or nonresident, who have a capacity to stand back from U.S. culture and reflect critically upon it. Applicants will be screened based on their knowledge, character, and personal skills. No formal academic qualifications are required (but see below for our additional screening of applicants who have received PhD's).

Duties

A national environmental leader must (1) help the people of the United States awaken to their environmental predicament, in its ecological and cultural complexity; (2) stimulate a yearning for better ways of dwelling on land and living with one another; and (3) insofar as possible, provoke the various cultural changes without which the country cannot achieve healthy lands and healthy people. The more particular tasks of the position, and warnings about hazards, appear in the material below. Given the nature of the underlying challenges, a long-term commitment is essential.

Qualifications

We seek a person of rare understanding, talent, and motivation; a person who is at once embedded in the modern age and yet experiences substantial detachment from it.

Our ideal candidate will present a background of eclectic personal experiences, in terms of interactions with people, nature, and serious thought. A wide-ranging curiosity is vital, as are an ability to reason clearly and a strong, even zealous commitment to the common good. The candidate will possess exceptional skills in communication, including the capacity to connect with varied audiences and a limitless patience in handling fools. The visionless, the impatient, and the readily depressed need not apply. A simmering anger and an inner frustration are acceptable, even desirable, if matched with a willingness to labor cheerfully against overwhelming odds.

A successful candidate must understand nature and its ecological and evolutionary processes, at varied spatial scales and over time. Detailed knowledge of a single scientific field is neither required nor particularly desired, and excellence in a single field will not excuse noticeable gaps in a candidate's breadth of understanding. A candidate's grasp of nature should feature evident strength in ecology, both as a body of knowledge and as a tool for gaining new knowledge. It must include, in addition to the basics of ecological interconnection and interdependence, an ability to talk sensibly to diverse audiences about (1) the limits on what we currently know about nature and (2) the recurring ways that scientific concepts, particularly about nature's overall functioning, are influenced by social and political values external to science.

Just as essential as knowing about nature will be a candidate's grasp of the human predicament, in its varied moral, social, cultural, and political aspects. We intend, on this general subject, to evaluate candidates carefully. Special attention will be paid to a candidate's ideas and sentiments on the following three topics.

Our environmental problems. Americans readily ignore the reality that our "environmental" problems are, at base, problems that relate to human behavior rather than problems inherent in the natural environment. Confusion on the point remains widespread, so much so as to provide a window on where American culture now stands.

Americans talk about environmental ills as if the planet itself were somehow to blame for them, rather than people and the ways we live. Our hope is that applicants in their submissions can propose particular ways to talk about nature and culture so as to highlight (forcefully but without insult) the deficiencies in the ways we dwell on land. Ideas are also welcome on ways to clarify the root causes of land degradation, within modern culture and within the institutions governing modern life (on which, more below). Here, too, Americans are in denial: They view instances of bad land use (polluted rivers, for instance) as discrete problems rather than as symptoms of deeper, more pervasive failings. The obvious fevers they admit, however reluctantly, but the underlying diseases they overlook. Their blindness here, naturally enough, is linked to their exuberant righteousness. Looking ahead, can Americans keep this moral confidence while becoming more aware of the harms that they cause to nature, one another, and future life? For the environmental leader, this question will linger, as a challenge and a hope.

Cultural criticism. Given how modern culture influences individual behavior, a successful candidate must firmly grasp the axial strands of modernity—the worldview, often traced to the Enlightenment, that so thoroughly pervades the era that we are essentially blind to it. Its major elements, as they relate to humans in nature, might be summed up in this way:

- Humans are distinct from other life forms; they are the sole possessors of moral value on the planet.
- Humans are best understood as autonomous individuals guided by reason and calculated self-interest.
- Individual humans possess rights, and their sole moral duty is to respect the rights of other individuals.
- Neither the dead nor the unborn possess rights, and thus neither enter into moral calculations.[1]
- Nature is mostly physical stuff that awaits our free use.
- Nature is complex and challenging to study (hence the Nobel prizes given out), yet ultimately it is no more than an intricate

mechanism; one day we will understand this mechanism and control it pretty fully.

- Our decisions about using this complex nature are rightly based on the knowledge we gain through empirical data collection; we need not include other forms of knowledge, nor should we adjust decisions to take into account our ignorance.
- A person's links to the places where she lives and works are entirely optional and easily severed; to no appreciable degree is a person constituted by that landscape, dependent on it, or properly defined in terms of it.
- In time, human cleverness can and will solve all problems.

A strong candidate for environmental leader must grasp the deficiencies (as well as the virtues) of these influential conceits. The ideal candidate, going further, should be able to awaken Americans to these shortcomings—helping people see how these embedded assumptions powerfully shape what they perceive, what they value, and even how they think. It hardly needs noting that this educational task will prove daunting. Americans embrace liberal individualism with passionate intensity. It is, in the common view, the appropriate if not inevitable end to moral progress (the "end of man"). Liberty, the key cultural value in this complex, is defined in such a way that only living individual humans possess it. This fragmented view of society-as-autonomous-individuals is strikingly similar to the prevailing view about nature. It, too, is viewed chiefly as a collection of parts rather than an integrated whole. Some of the parts are valuable (natural resources); most of them are not. All or nearly all are suitable objects of private ownership.

These prevailing ideas sink deep roots within the "free" market—the institution that now dominates and defines American society. No liberal institution has ever wielded such power. For the environmental leader, the market will be an occasional friend and a constant, well-armed opponent. In the market and thus in America generally, value is established chiefly by the purchasing decisions of people with money. Because of this, only living people count for anything. This is so in practical terms, and thus it is a moral truth.

Nature enjoys value only when humans pay to protect it; future generations are respected only in the same way.

Challenging and countering these cultural assumptions will make up a good part of the job of the environmental leader. In some way, the leader must promote cultural ideals that recognize the ways humans are integrated into nature and with one another. The leader must instill ways of understanding that admit human ignorance and the limits on reason while embracing broader moral values and showing respect for the long term.

Good land use. Environmental ills are perhaps best understood simply as forms of bad land use (so long as we define "land" broadly to include all of nature's components operating as an integrated system). To speak of bad land use, however, is to presume some standard that distinguishes the good from the bad. At the moment we have no such standard, and few Americans, conservation leaders included, have thought seriously about it. An ideal environmental leader will have done so and should be able to help others think about it as well.

For decades now, arguments have taken place over which tools we should use in promoting environmental progress: regulation, economic incentives, disclosure requirements, trade sanctions, publicly funded research, hard-look study rules, new decision-making processes, and the like. Meanwhile, hardly anyone takes time to define environmental progress—the overall goal—other than in fragmented ways (clean air, clean water, protected species). If conservation succeeded overall, what would success look like? Vague terms are tossed out for consideration: sustainability, ecosystem management, sustaining ecosystem services. But they capture only part of the story. Yes, we need to plan for the long term (sustainability). Yes, we need to think about broad spatial scales (ecosystem management). And yes, we need to respect nature's fundamental ecological processes or "services" (the most useful of the three terms). But there is more to it.

Good land use would take into account the full range of human needs, including aesthetics and the comforts of living in sensibly arranged communities. A constellation of ethical and religious considerations are also relevant, particularly ones relating to future generations and rare species. In some way good land use would pay attention to our ignorance about nature and to the wisdom of acting cautiously in the face of that ignorance, avoiding bets that we cannot afford to lose. Merely to scratch the surface of these many factors relevant to good land use is to bump repeatedly into nature and into the wisdom of respecting nature's ways. Good science is required, of course, to make many of these decisions, but science rarely provides answers on its own. It must be brought together with an array of normative considerations, and the work of bringing them together is not easy.

On this key issue—distinguishing land use from land abuse—an environmental leader will confront an especially discomforting reality. The U.S. environmental movement is highly fragmented. Groups compete more than they cooperate, and they rarely support unified messages. The movement as a whole has no overall goal, and apparently no way even to formulate such a goal. With the conservation movement so voiceless, it is easy for critics to accuse the movement of misanthropy and other vices. Thus many citizens assume (wrongly) that conservation means favoring nature over people; they are unaware that people are embedded in nature and ultimately dependent on nature's healthy functioning. For the environmental leader, the difficulties here will be vast. Environmental groups of course can be allies, but absent a sea change in institutional cultures, the groups will be slow to work in concert.

The selection process for this position will turn on the above factors and on a candidate's understanding of the major institutional components of modern society. We identify five of these components because of their critical roles in shaping the ways people relate to nature. An ideal candidate will show deep understandings of these institutions and will propose ways, as appropriate, to reform them and to keep them in their proper places.

The free market. Foremost—as noted—is the free market, which orchestrates most resource-use practices of Americans. So immersed are Americans in the market that they rarely think about what it is and what it can do. Market failures are poorly identified, particularly the market's inability to promote sensible land-use practices. Unleash the market, and all will be well—so it is said and so many Americans believe. Turn all of nature into private property, and the market's invisible hand will promote conservation while shifting resources to their highest and best uses. This refrain, too, is often heard. Just enough truth resides in these contentions to give them social traction. Necessarily, sound environmental policy would keep the market within limits, to protect nature's functioning and to fulfill our felt ethical duties to future generations, other life, and one another. Alas, we are a long way from having such limits in place. We are even far from knowing what the limits ought to be.

Science. Americans are less emotionally attached to science, yet it too enjoys respect, and they are equally poor in understanding what it can and cannot do. The call has long gone out to base our environmental policies on "sound science." We utter this position without recognizing that science alone is unable to distinguish good land uses from bad ones. Science provides only bits and pieces of what we require to assemble a vision of good land use. Particularly muddled is our thinking about a critical moral question: how much evidence should we require about a potential environmental problem before deciding to take action? In the case of potential terrorist threats, we act on evidence that is slight and unreliable. Why then should we wait until an environmental ill is scientifically "proven" (as some claim we should) in the sense of being supported at a confidence level of at least 95 percent? And why limit our consideration, when we judge dangers, to evidence that itself qualifies as "scientific"? We impose no such requirement in any other area of public policy, and it makes no sense here. Is it wise to sit back and do nothing simply because a problem is only 80 percent likely to happen? Should we wait before acting until scientists take note of facts that are evident to ordinary people? How are the huge gaps in what science knows taken into account?

For the environmental leader, hardly any constellation of problems will be more knotty than those having to do with science and its rightful roles. Of course good science is better than bad science, so we should insist that science be as good as reasonably possible. But it rarely makes sense to require scientific proof before acting, nor is it wise to insist that people who care about the environment always bear the burden of proof. All relevant evidence should be weighed, whether or not it has been vetted in peer-reviewed studies. Too often the call to give science a bigger role is merely a huge wrench thrown into the gears of environmental policy making to cause a breakdown. The demand for more science brings delays and inaction. It can raise high if not impossible barriers to sensible solutions. It can sow doubt where very little exists, while challenging the integrity of citizens who warn of environmental threats. Worst of all, it can turn fundamental policy questions into seemingly technical issues in ways that disarm and disenfranchise concerned citizens.

Public governance. Along with the market and science there is the whole matter of public governance and the disturbing lack of good mechanisms for citizens collectively to formulate sensible land-use goals. Democracy in America is sick, particularly when it comes to decision making about our shared natural homes. Landscapes remain ecologically integrated no matter what we do to them. That reality does not change when we fragment the landscapes among landowners and political jurisdictions. Many land-use goals are achievable only when plans are prepared on a landscape or watershed scale. If good land use is to come about, collective action is thus essential, which means democratic engagement is essential. An untiring environmental leader is needed to help bring it about.

The challenge here, in terms of decision-making mechanisms, goes well beyond reviving democracy, sapping power from big money, and getting power back to citizens. New mechanisms are urgently needed so that people can gather and give thought to their shared landscapes and to ways of making them better. Study and deliberation are essential; it is not enough to rely on responses to public opinion surveys. Sound governance mechanisms would pay attention to nature's own lines—particularly its rivers and watersheds—more than to the arbitrary lines that people draw on maps.

Inevitably, many environmental challenges will require study and action at varying spatial scales, from local to national to global. This means there is need to allocate power among levels of government. What power should go to what level? And what happens when one level of government fails to do its job? A well-prepared environmental leader will offer answers or at least stimulate good thought.

Higher education. All of these tasks would be easier if America could look for help to its institutions of higher education. But higher education has become as much a part of the problem as it is a solution. Academic knowledge is ever more fragmented. Fewer scholars can grasp the whole of things or sensibly assemble the many intellectual pieces. In truth, the market's tentacles have replaced the ivy on our university towers. Today's academy helps design market products and prepares students to be good producers and consumers. Meanwhile, it gives diminishing thought to reforming society and producing good citizens. Academic research is largely about generating new facts and building technology, not about probing normative issues, tracing problems to cultural flaws, or painstakingly distilling diffuse knowledge into usable wisdom. Colleges train students to be inhabitants of the world, which is to say inhabitants of no place in particular.

Universities are hardly more aware than citizens generally that our environmental problems have to do with behavior and culture. At the typical institution, environmental studies is merely a collection of applied sciences with a few economists and policy people tossed in. The group is little more than the sum of its narrowly focused parts. Few programs recognize that sound environmental thought takes place at higher levels of synthesis and integration. An ideal environmental leader should thus be forewarned: Do not look for peers within the academy (though useful allies can be found). Should you reach out to the academy, the likely response will be disdain for your alleged lack of disciplinary rigor. In today's academic pecking order, status is reserved for the high-yielding specialist. Nay, matters are even worse. It is reserved for the specialist who obtains and spends the most research money. Research output is important chiefly insofar as it leads to higher monetary inputs.

For these reasons, we must make the following special note for candidates with PhD's. Although they are eligible to apply for this position, they must in their applications provide evidence that they are able, despite their training, to think broadly about issues of nature and culture. They must give evidence of a capacity to make full use of varied disciplinary perspectives. In particular, holders of doctorates in scientific fields must include with their applications an essay that (1) identifies and evaluates the limits of science and scientific research models and (2) explains why additional scientific research is unlikely to solve most environmental ills. Holders of doctorates in economics must include a detailed critique of the assumptions of neoclassical economics and explain clearly why and how contemporary economic thought has exacerbated the major flaws in contemporary culture.

Private property. The final special challenge that an environmental leader will face is the institution of private property. Americans hold it particularly dear, and confusion over this institution is hardly less than it is with the market and science. Scarcely anyone seems to understand how private property works, in terms of using government power to restrict the liberties of people other than the landowner. Americans assume that private property means being able to use your land any way you like with little regard for consequences. This is a far cry from where landownership stood two centuries ago, when sensitive land uses were protected against interference by noisy neighbors. Private property is the most important institutional mechanism that allocates decision-making power over land. Bad land use is often the result of bad decisions of private owners. The institution cries out for reform.

In one sense, though, it is not private property that is the real problem. Private ownership is a flexible arrangement. Landowner rights can be redefined by law in ways that require owners to act responsibly. When sensitively structured, the institution can become a good tool to promote healthy lands. The true problem, then, resides deeper. It resides, as usual, in the minds and guts of Americans. Failing to understand the institution, we have essentially lost control of it. We do not know how it works, nor do we realize the many policy choices embedded in it. So far as most people can tell, the conflict today is about whether we will or will not protect private ownership rights securely. The truth is quite otherwise. This issue is not whether we will protect property but how will we define private property, in terms of the rights and responsibilities of owners. Our blindness on this critical issue goes far toward explaining why land-use democracy is so weak. When we instinctively hand managerial power over to private owners, leaving them largely free to act as they like, what power is then left for the demos?

Working Conditions

Many Americans are prepared to help a national environmental leader. Indeed, the vast majority are likely to do so. That is the good news. But first a leader needs to get their attention, and that will prove hard. The institutional obstacles noted above—love of the market, confusion over science, weak democracy, diverted academies, inflexible property regimes—are all deeply entrenched. To this mix we should add a news industry that mischaracterizes environmental issues (in the name of making them understandable), a constant flow of corporate greenwashing, and pro-industry political groups that deliberately sow confusion. Mix these together, and we have the makings of a true mess.

Dealing with this cultural mess will be the environmental leader's prime job. An able leader naturally will be alert for allies. And they are out there, waiting to be found. Many of them will come from religious groups who are ready to hear about the ways that land health is a moral good. Other allies will reside on small farms and in settled neighborhoods. They are ready to rise up when told about the ways that citizens can wrest control over their natural homes from developers and industrial land users. Deep down, many Americans sense that scientists know less than they claim. These citizens, too, could respond when they realize that much of what passes for science goes well beyond it. All Americans respect private property, and rightfully so. What they await (though they may not know it) is news about how they can respect private property while insisting that landowners behave in responsible ways. They simply need to realize that they can have it all.

Finally, there is the largest of all groups of potential allies. They are the citizens who care about nature but who have trouble imagining how things might be different. Surrounding landscapes, they sense, could be far better than they are. But what could they look like? If we lived well on land, how would we be living? More than anything, Americans need a vision of a better way, a way that respects nature and keeps it close at hand; a way that honors health above wealth and that prepares for generations unborn. An ideal environmental leader, then, would hold high a torch, not to identify that better way precisely but to illuminate the path that can lead toward it. We need strong light, and soon.

Timetable

The application period for this position is unlikely to close, and applications will be reviewed as received. In the event one environmental leader is found, the search will continue for others, perhaps without end.

Note

1. One early challenge that an environmental leader will confront is the confused rhetoric used by opponents of legalized abortion, who employ the label "pro-life" yet unashamedly define "life" so as to exclude nearly all forms of life found on the planet. They show little or no interest in respecting the essential processes upon which all life depends, people included. That this use of the term "life" goes unquestioned in contemporary politics provides evidence of the firm grip of Enlightenment thought and highlights the cultural barriers to the emergence of a true agriculture.

9

The Politics of Homeland Health

One lesson that stands out from the past few federal elections is that the land-conservation stance is due for an overhaul. Fifteen years ago, the first President Bush matter-of-factly declared himself an environmentalist. Today, few Republicans and a good many Democrats would accept the label only with qualifications, if at all. In the 2004 presidential election, John Kerry and other pro-environment candidates had little or nothing to say about the subject, presumably for fear that strong comments could cost votes. Few voters today treat the environment as a decisive issue. Something has happened.

Substantial credit for this anti-green shift goes to the capable and well-funded environmental opposition, which has been aided by journalists either unwilling or unable to navigate the polemical minefields. But just as much credit (or blame) belongs to the environmental movement itself, which is fragmented, incoherent, and largely unable to defend itself against opposing claims. In the squabbles over culture, symbols, and slogans, the environmental cause is getting beaten. People are confused about what the cause seeks to accomplish and about the resulting costs. They worry also about the effects that environmental rules would have on liberty, private property, and the nation's ability to compete internationally.

Things need to turn around, for the sake of the land and of progressive causes generally. And they can if people who care about land and good uses of it deal with their internal problems, identify why they have stumbled, and then craft coherent, morally compelling ways of addressing our environmental plight. The cause needs to stress new themes: ensuring the health of the land community as a whole, protecting life and creation, and investing in America's future in ways that yield big dividends. It needs to rise above squabbles over science and burdens of proof while directly challenging the allegations of high cost. Most of all, it needs to talk consistently and forcefully about good citizenship, future generations, and the morality of living responsibly.

Environmentalism these days is widely deemed a liberal cause. The label is apt in the sense of affirming a willingness to use government to promote the common good, but not in the classic sense of liberalism as a move to liberate individuals from limitations on their pursuit of self-interest. Environmentalism promotes human health but has little to do directly with personal autonomy. It is better viewed as a communitarian perspective in that it promotes the healthy, long-term functioning of communities as wholes, nature included. Although environmental leaders may talk about an individual right to a healthy environment, the central strand of environmental thought focuses on nature and how humans ought to live in it. Among communitarian causes, environmentalism stands out because it defines the relevant community broadly, to include other life forms, future generations, and even the geophysical earth itself, linked in webs of interdependence. If better understood, the cause would enlist support from across the political spectrum.

Environmentalism grew out of the land conservation cause, which was led during the middle decades of the last century chiefly by businesspeople. Typical of them was Republican Horace Albright—zealous advocate for national parks from 1914 until the early 1960s, sometime head of the National Park Service, personal friend of Herbert Hoover, and between 1933 and 1956, while vigorously promoting nature conservation, general manager of United Potash Company. In recent decades, congressional Democrats have consistently compiled stronger environmental voting records than Republicans. Only within the past decade or so, though, has this voting gap become a chasm. As recently as the late 1980s, Democrats received ratings in the 65–70 percent range from the League of Conservation Voters; Republican ratings hovered near 35 percent. Since then the margin has nearly doubled. Democratic ratings have risen above 75 percent; Republican ratings have fallen to approximately 13 percent.

The rise in Democratic support has come mostly because legislative votes have dealt with measures to curtail environmental programs, not extend them. In all likelihood, Democratic support for environmentalism has softened. More politically significant is the drop in Republican support, which has come chiefly because the party's economic libertarians, not its social conservatives, have defined the party's stance. Republican leaders portray the environment as an issue relating to free enterprise and American competitiveness, not to morality and responsible living. This is where the shift has occurred, and it is where the counterattack should aim. The environment needs to regain its status as a moral issue, as a question of right and wrong, tied to religion, distinctly pro-life, and linked to the welfare of generations to come. Voters must find homeland health just as compelling as homeland security.

Despite current headwinds, land conservation can again become an issue that unites voters who otherwise disagree. Economic libertarians are unlikely to warm to it, but social conservatives certainly could. No conservation author stands higher among today's faithful than Wendell Berry of Kentucky, whose morally charged writings reiterate conservative themes of the New Humanists and Southern Agrarians of the 1930s. The late Richard Weaver, conservative icon and author of the postwar classic *Ideas Have Consequences*, could comfortably fit within today's land conservation cause. If well constructed and linked to core American values, a green message about caring for land and life ought to appeal to old-style conservatives who place morality and community above the capitalist market.

To turn matters around, environmental advocates need to work on both their offense and their defense. They need to rethink what they are about, what they are trying to accomplish, and the values and principles they hold high. Just as important, they need to identify why voters are so worried and why the cause has stumbled. The rhetoric used by opponents is not overtly anti-environmental. It is positive rhetoric, phrased in terms of science, economics, liberty, private property, and prosperity. Environmentalists need to address these matters head-on—without hiding behind posters of wolves, caribou, and puffing smokestacks. It is not enough to talk about specific controversies and program initiatives, nor is it adequate to use vague slogans such as "connecting people to nature" or "conserving land for people." Morality needs to be part of our message, and so does an ecological understanding of the human predicament.

An initial challenge in redirecting conservation comes from the bad connotations that opponents have attached to the terms "environment" and "environmentalism." These used to be good words with strong associations. Now they are ambiguous at best. Many green groups know this and use the *e* word less often, if at all. They are conservation groups, stewardship alliances, or groups that promote wildlife or sustainable agriculture. A growing number of them are land trusts, which steer as far from the *e* word as they can. The problem with the term is that many audiences link it to by-now-familiar complaints about the cause. The cause is elitist, critics assert, especially when it protects playgrounds and scenic vistas for the rich. It is also misanthropic, caring about nature more than people. Regulatory measures trample on private property rights and infringe individual liberties. The cause costs jobs and undercuts business competitiveness. The litany of complaints is familiar.

Terminology aside, the starting point in the rebuilding effort is to realize that voters continue to support environmental goals, and by wide margins. What makes them nervous are the means used to achieve the goals—the regulations, land-use planning, permit systems, and restrictions on property rights. Critics know this full well, which is why they swing hard at the means, attacking them ceaselessly as expensive, intrusive, and un-American.

There is no need to abandon the term "environmentalism," but the cause needs new phrases and strategies to confront its adversaries. Advocates need to focus on the overall ends of environmental policy and on the basic values at stake, not the means. What are we trying to accomplish, and why? What are the values we are trying to promote? On questions about the means, the wise approach is to endorse pragmatism and support any tool that works, including market-based measures. Means are unimportant so long as they work well and fairly for citizens, landowners, and taxpayers.

Another reason to avoid dwelling openly on means is that debates quickly bog down in nasty factual disputes. Environmental problems are more complicated than they used to be: more ecological, more indirect, longer term, and harder to spot. Opposing interests are skilled in producing studies that appear to challenge nearly every factual claim—for instance, that farm chemical runoff is a major cause of Gulf hypoxia or that bioengineered crops pose genetic threats. Few audiences have the attention span required to wade through the allegations. The bottom line: the more the discussion stays centered on values and overall goals, the more effective the pro-land message will be.

As for goals, it makes sense for environmentalists to avoid the imminent-health-threat line of argument. The rhetoric does not fit many of today's issues, people are weary of it, and relevant facts are complicated. Instead of talking negatively about risk and danger, they should speak positively about health. The environmental cause seeks to secure healthy lands and healthy people. It promotes care for the earth for future generations. It upholds the quality of life of people living today. These are the points to emphasize.

We can learn here from current discussions about Social Security and Medicare. These programs do not face immediate crises, and yet the public is worried about them. Both systems are slowly failing, or so it is said. In the case of the environment, the message should be the same. The land as a fertile, ecological system is on the decline, and it is taking many life forms and human options along with it. Global climate change supplies a clear example of this decline, as do the downward fates of many wild species. Our ways of living in nature need correcting, before problems get even worse.

When talking about overall aims, environmental supporters ought to refer routinely to the welfare of the entire community of life, humans included. They should talk about the land as an integrated, ecologically functioning system, rather than straining to translate every environmental ill to the individual level. It is this integrated, organic system that needs to stay healthy if humans are going to prosper. Only in the case of disputes over specific forms of contamination (such as mercury emissions from power plants) is it wise to reduce matters to the level of direct health threats to individuals.

It also makes sense to avoid openly fretting about looming shortages of natural resources—the other familiar way of talking about environmental ills. In truth, declining stocks of many natural resources (metals, for instance) do not pose direct environmental problems. Nature's functioning is not declining ecologically just because we are using up underground supplies of copper or iron ore. Indeed, declining oil supplies can be a positive development ecologically. When it comes to oil—which is harmful both in extraction and use—our chief environmental worry is not that we are running out but that we are not running out fast enough. For most dwindling resources, the market is likely to find reasonable substitutes. What the market cannot and will not do is keep the land healthy in an ecological sense. It will not, on its own, produce landscapes that are healthy and pleasing for people, ecologically, aesthetically, and ethically. In any event, to talk about nature as a warehouse of resources (as the resource-shortage argument does) clashes with the more integrated, ecological view of nature that needs to stay front and center in all environmental rhetoric. Beyond that, such talk needlessly challenges America's faith in the market, is easily attacked with anecdotes (did we not worry 150 years ago about running out of whale oil?), gives the whole cause an unnecessarily negative cast, and unwisely questions America's ingenuity.

Our chief worry, in short, is neither about resource shortages nor about direct health threats to people living today, important as they are. It is about the healthy functioning of the land as a whole, as an integrated natural system that includes people. We depend on the healthy functioning of this interconnected land. So will our grandchildren and their descendants. So too do all other forms of life, from the soaring eagles and lumbering grizzlies to the flittering butterflies and thrashing salmon. Environmentalism is about land health.

An environmental message focused on health, for land and for people, has obvious moral overtones. Consistently and firmly used, the message could help the environmental effort gain moral ground. So too would other rhetorical tools that link environmentalism to living responsibly. Environmentalism is about showing compassion and concern—for neighbors, for one's community, for children and future generations, for other life forms, and for creation as a whole. Rhetoric about responsible living has particular value in responding to the recurring criticisms. Yes, private property is an essential element of American society, but surely we can insist that landowners act responsibly. Yes, liberty is also essential, but what Americans have always valued is the liberty to live responsible lives. The opposite of responsibility is not freedom; it is license.

Among the terms opponents have deployed in efforts against environmentalists is the term "voluntary." Conservation ought to be voluntary, critics claim, not coercive. Voluntary action is more virtuous; voluntary conservation shows proper respect for individualism and private property rights. It is a positive word, resonant of liberty, linked to the American way, and useful both in resisting regulations and in justifying cash payments to landowners who refrain from harmful development.

The environmental cause has developed virtually no response to this stance. To make conservation voluntary, we ought to be insisting, is to defend irresponsible conduct. It is to shield landowners and polluters who want to keep causing harm and do not like being told to stop. In the case of property rights, the voluntary approach protects landowners who assert the right to misuse their lands—often harming their landowning neighbors. A true volunteer is someone who practices conservation for free, not someone who gets paid to do so under contract. And when government does the paying, we need to consider the taxpayers, whose tax payments are certainly not voluntary. The real dichotomy, then, is not between voluntary and coercive conservation; it is between conservation burdens imposed on landowners, who largely cause the problems, and burdens imposed on taxpayers, who usually have nothing to do with them.

To regain moral ground, the environmental cause might also make effective use of the expression "pro-life." The term requires careful handling, to be sure, yet its possibilities are alluring. Does not a pro-life position logically include clean air and clean water? What about the thousands of Americans who die annually from air pollution emitted by power plants? Can a person logically carry signs demanding that a Terri Schiavo keep her feeding tubes while protecting and expanding Louisiana's cancer alley? Environmental advocates might similarly employ the term "conservative" to remind audiences of its verbal link to "conservation." Once upon a time, the two terms were used interchangeably. They could be again. Land use is conservative when it respects the land's functioning. True conservatives respect the community of life.

The environmental cause needs to do a better job dealing with the opposition's claims. Some of the allegations are obvious; for example, that environmental protections undercut property rights. Other are not so obvious yet do just as much damage.

On top of the not-so-obvious list is a set of issues having do with the factual and scientific groundings of environmental claims. Solid facts are essential, and everyone can see that environmentalists and their opponents wrangle about them. What is less clear and more damaging are the fights over who should be required to present proof, what the burden of proof ought to be, and what qualifies as evidence. The issues seem mundane if not technical, but they are hugely important.

The environmental resistance has learned key lessons since the early 1960s, when it unleashed its furious frontal assault against Rachel Carson. Opponents learned that they do not have to win the scientific debates about environmental decline. It is up to the environmental side to prove its case to the public; opponents need only to sow seeds of doubt. To do that while calling for more scientific research is to take a stance more reasonable and appealing than direct opposition. (We see the approach used effectively on global climate change, for instance, and on issues about endangered species and toxic emissions.) Opponents also learned how to capitalize on America's respect for science. Americans respond favorably to the call to base public policies on good science because the stance sounds sensible. Meanwhile, few people recognize how much it favors the environmental opposition, given that scientific "proof" requires far more than the courtroom's preponderance-of-the-evidence standard (the former is often pegged at the 95 percent confidence level). What sense does it make, though, to ignore a problem or danger simply because we cannot prove scientifically that it is taking place? Do we ignore serious problems just because they are only 80 percent likely, or 50 percent, or even 20 percent? And as the research proceeds, of course, nothing changes.

The environmental cause needs to steer clear of litigating facts in the public arena because the public will not pay attention. The better strategy is to focus on basic values and on the elements of practical wisdom. There is plenty of evidence that we should take environmental worries seriously, and that is just the way to talk about it. We do not insist on scientific proof in international affairs or in everyday life, nor do we base decisions only on data gathered in the course of scientific studies. The same pragmatic approach should govern our dealings with nature. Environmental protection is a kind of social insurance policy. It is about avoiding unsafe conditions and needless risks.

Probably most troubling of all, on this issue of contested facts and burdens of proof, is the subtle move by the Christian right to encourage Christian voters to ignore science altogether on planetary issues. If half of all adults can deny evolution, why not also deny the possibility of global climate change or resource shortages? If God created the earth in a week, giving it to humans to tend, surely he can save the earth from any catastrophe (or save the faithful, at least). After all, did not God provide for the Israelites in the desert when their natural resources ran out? It is difficult to gauge the political effect of books such as the best-selling Left Behind series, but a literal embrace of their fantastic arguments could lead a person to ignore even well-grounded warnings of ecological disaster. Who knows but that environmental destruction might foretell a looming day of judgment, the advent of the much-desired end-time.

Just as damaging for the environmental cause as these science traps are the frequent claims that environmental protection is simply too expensive or that it undercuts America's competitiveness. Evidence supporting these claims is exceptionally weak, and the land-conservation cause needs to take the offensive on the issue. Plentiful evidence points in precisely the opposite direction, that environmental controls bring economic gains rather than net costs. States with the toughest environmental standards typically have the highest wage rates.

Post hoc cost-benefit assessments of environmental regulations (excluding occupational safety standards, which pose different issues) routinely show economic benefits that exceed total costs. According to a 2002 Office of Management and Budget report, the estimated range of annual costs associated with all major federal environmental regulations had a midpoint of $162 billion; the estimated range of annual benefits had a midpoint of $999 billion. Utility industry estimates in 1989 placed the costs of the Clean Air Act's then-new sulfur dioxide program at between $4.1 billion and $7.4 billion per year; by 2000, actual costs were estimated at from $1 billion to $2 billion per year, with benefits being perhaps ten times higher. Industry estimated in 1990 that it would cost $14.8 billion per year to comply with new pollution standards for volatile organic compounds emitted by stationary pollution sources; the actual costs as of 2000 were pegged at about $960 million. In January 2005, a report written by a Harvard scientist, commissioned by the Bush administration's Environmental Protection Agency and peer reviewed by its staff, calculated the overall benefits that would come by strictly reducing mercury emissions from power plants (largely a matter of enforcing current law). When, two months later, the EPA publicly announced why it rejected the strict guidelines, it ignored the Harvard study and set the estimated benefit level 99 percent lower.

When we get beyond pollution-related squabbles, where the cost-benefit facts are the most confusing, the economic benefits of conservation become undeniable. Energy conservation can produce big savings compared with the costs of higher energy generation. Compact communities, with their lower infrastructure and transportation costs, are big savers. New York has learned that it is far cheaper to protect watersheds in the Catskills than to treat drinking water once it is polluted. Communities along the Mississippi River have learned that it is cheaper to protect wetlands upstream than it is to build levees and berms to protect against flooding. The list goes on: conservation routinely saves.

The confusion on cost issues arises mostly because critics have done such a good job narrowing inquiries so that they overlook or discount most economic gains. We need to be candid: conservation laws do produce losers as well as winners. Real people lose jobs, and real companies incur net costs. But this is simply economic dislocation, of the type that the market produces every day (and that is trivial in comparison to corporations' moving factories overseas). Job losses in one place are often offset by job gains elsewhere. Finally, there are the recurring complaints that environmental measures are costly because they interfere with private property rights, undercut individual liberty, and are otherwise inconsistent with core American values. Here again, the environmental cause needs to get its act together and respond.

Opponents typically present allegations about invaded property rights in anecdotal form. Hapless landowners who simply want to use their lands in ordinary ways are told they cannot. The anecdotes fail to mention neighbors and other nearby residents, who have their own property rights and who might well be harmed by what the hapless landowners want to do. To stop one landowner from draining his wetland is to protect another landowner against flooding. A good way to deflect anecdotes is to bring back the missing characters—the neighbors, the people downwind, and so on. When the missing characters and their interests are also considered, environmental laws are more likely to promote property rights and individual liberty than to undercut them.

The main defense, though, is this: Environmental laws protect responsible, conservative landowner activities. They call upon owners to abide by the age-old principle of doing no harm, the bedrock rule of owning and using land. They also call upon landowners to do their fair, reasonable shares in upholding the healthy functioning of their home landscapes. Environmental laws protect the downstream landowner from being flooded out when upstream owners drain their wetlands. They protect the rural dweller from having her home rendered unlivable by the stench of a nearby hog operation. Future generations will inhabit the same lands one day; should we not show respect for them? Countless citizens enjoy fishing, hunting, swimming, and watching birds; how do their liberties fit into the story?

The conservation effort needs to take the initiative on issues of economics, liberty, private property, and others. The true misanthropists are those who pollute our air and drinking water, not those who try to keep it clean. And what forms of recreation could be cheaper and less elitist than camping, fishing, hunting, and simply enjoying the beauties of nature?

Since it began in the 1920s and 1930s, the organized conservation movement has been characterized by countless organizations with varied aims and methods. Many environmental professionals exalt this diversity: conservation is a big tent, they announce, with room enough for rich and poor, urban and rural, black and white. That may be so. But the cacophony that goes on within the tent comes at high cost. Lacking cohesion, the movement has few or no acknowledged spokespeople. It has no organized ways to counter widespread charges and to talk about economics, liberty, and private property. Meanwhile, political candidates have no good ways to discuss the environment because there are no messages, already familiar to voters, that they can weave into their speeches.

Things need to change, and that can happen only if environmental leaders get together, take stock of where they are, develop common stances and rhetoric, and then work together to present shared messages. Too often groups now work at cross purposes, with land trusts and some of the biggest organizations content to work within the system, challenging nobody and asking only for enough money to buy some pieces of land here and there. In the meantime, they undercut the work of the true reformers, those who realize that American culture is defective and that business as usual in the land-use game will gradually pull down all lands everywhere.

New messages, carefully phrased and consistently repeated,

could push voters to see the link between environmental disputes and responsible living. Environmental protection is about promoting the health and prosperity of the entire land community. It is about protecting life, caring for creation, and investing in America's future in ways that yield big dividends. It is about acting cautiously in the face of dangers, looking out for our grandchildren, and preserving the simple joys of living in nature. Most of all it is about good citizenship, responsible living, and following the moral high path.

ACKNOWLEDGMENTS

Several of the chapters of this book were delivered originally as addresses to diverse audiences, and I am grateful to the sponsors of the various gatherings for letting me air my views. Chapter 1 was presented in October 2004 as an address to the international Natural Areas Association's annual gathering in Chicago, Illinois; it then carried the title "Natural Areas in Place and Time." Chapter 2 began life even earlier. I presented a version at the conference Wilderness Britain? held in Leeds, England, in March 2001, under the title "Aldo Leopold, Wilderness, and the Quest for Land Health." Chapter 5 was offered several times in evolving forms, in January 2005 to the annual conference of the Quivira Coalition in Albuquerque, New Mexico; in March 2005 to a gathering at the Wallace Stegner Center at the University of Utah College of Law; and (in close to its current form) in September 2005 to the students and faculty at Lewis and Clark Law School in Portland, Oregon, where I was honored to serve as the school's annual distinguished natural resources law visitor. The talk was published in the school's journal as "Good-bye to the Public-Private Divide," *Environmental Law* 36 (2006): 7–24. Chapter 6 was delivered in January 2004 under the title "Ecology, Ethics, and Private Land" in a campus-wide lecture series on agriculture and food issues at the University of Minnesota in St. Paul.

Chapter 9 was written at the request of the editors of the political journal *Dissent* as a commentary on environmental politics;

it appeared as "Homeland Health: How Environmentalism Can Regain Lost Ground," *Dissent* (Summer 2005): 48–53. My thanks go to Jim Rule of *Dissent* for suggesting the topic and providing guidance on my drafts. In the case of many chapters of this book, I benefited from extended discussions with Julianne Lutz Newton, and I am especially grateful to her.

At the University Press of Kentucky, series editor Norman Wirzba approached me about my work and kindly invited me to contribute to his agrarian series, and press director Steve Wrinn has been unfailing in his support and professionalism; I thank them both. Scott Russell Sanders served as an outside reviewer and gave useful comments for trimming and focusing the work. I thank him as well an anonymous reviewer whose suggestions also proved sound.

INDEX